The Diffusion of Military Power

The Diffusion of Military Power

CAUSES AND CONSEQUENCES FOR INTERNATIONAL POLITICS

Michael C. Horowitz

PRINCETON UNIVERSITY PRESS

PRINCETON AND OXFORD

Published by Princeton University Press, 41 William Street, Princeton, New Jersey 08540

In the United Kingdom: Princeton University Press, 6 Oxford Street, Woodstock, Oxfordshire OX20 1TW

Library of Congress Cataloging-in-Publication Data

Horowitz, Michael, 1978–

The diffusion of military power : causes and consequences for international politics / Michael C. Horowitz.

p. cm.

Includes bibliographical references and index.

ISBN 978-0-691-14395-8 (hardcover : alk. paper) — ISBN 978-0-691-14396-5 (pbk. : alk. paper) 1. Military art and science—Technological innovations—History. 2. Military art and science—Technological innovations—Political aspects. 3. Military art and science—Technological innovations—Economic aspects. 4. Balance of power. 5. International relations. 6. Military policy. I. Title.

U42.H67 2010

355'.07—dc22   2009043474

British Library Cataloging-in-Publication Data is available

This book has been composed in Adobe Caslon Pro

Printed on acid-free paper. ∞

press.princeton.edu

Printed in the United States of America

1 3 5 7 9 10 8 6 4 2

CONTENTS

ILLUSTRATIONS

FIGURES

TABLES

This book examines the spread of military power throughout the international system, explaining how variations in the diffusion of new military innovations influence international politics, especially the balance of power and warfare. States have a number of possible strategic choices when faced with military innovations, including adoption, offsetting or countering, forming alliances, and shifting toward neutrality. My theory, named *adoption-capacity theory*, argues that for any given innovation, the financial resources and organizational changes required for adoption govern the system-level distribution of responses and influence the choices of individual states.

As the cost per unit of the technological components of a military innovation increases and fewer commercial applications exist, the rate of adoption decreases and alternatives like forming alliances become more attractive. Similarly, if implementing an innovation requires large-scale organizational changes in recruitment, training, and war-fighting doctrine, fewer actors are likely to adopt it. While higher financial requirements generally mean adoption patterns will benefit preexisting wealthy and powerful states, however, higher organizational change requirements can handicap the wealthiest states, and upset the balance of power toward smaller and more nimble actors.

Using multiple methods ranging from large-n statistical tests to the in-depth analysis of primary sources, I test the theory on four cases: nuclear weapons, battlefleet warfare, carrier warfare, and suicide bombing. The results strongly support the theory, and the suicide-bombing case demonstrates its conceptual reach beyond state military organizations to explain a key trend in international politics. This chapter views suicide bombing as an innovation, and discusses how financial and organizational constraints influence terrorist groups' decisions. For example, the high organizational change requirements for adoption explain why older, previously successful terrorist groups like the Provisional Irish Republican Army (PIRA) and the Basque Fatherland and Freedom Group (ETA) did not adopt suicide terrorism, but Al Qaeda did. The conclusion moves forward and examines the way potential information age shifts in the production of military power could influence the future of the international security environment for both state and nonstate actors, including the United States, China, and Al Qaeda.

I have been fortunate to receive tremendous support at every step along the way in the long process of writing this book, incurring a great number of intellectual debts. While any errors in this work are most certainly mine and mine alone, there would definitely have been many more without the help of a great number of people.

I was privileged to work with an exceptional group of senior scholars while in graduate school, including Stephen P. Rosen, Alaistar Iain Johnston, and Allan Stam. Each made a vital contribution to the ideas and arguments in this book. I would also like to thank some of the many other people who took time out of their own research to read and comment on pieces of my manuscript—or the whole thing in some cases: Nikolaos Biziouras, Bear Braumoeller, Jeff Frieden, Michael Glosny, Stephanie Kaplan, Andy Kennedy, Andy Kydd, Paul MacDonald, Manjari Miller, Assaf Moghadam, Phil Potter, Daryl Press, Sebastian Rosato, John Schussler, Beth Simmons, Erin Simpson, Robert Traeger, Steve Walt, and Catharina Wrede-Braden.

Financial support from the Government Department at Harvard University, the Weatherhead Center for International Affairs, the Olin Institute for Strategic Studies, and the Belfer Center for Science and International Affairs gave me the time and opportunity to complete the early stages of my research. Steve Bloomfield in particular helped make the Weatherhead Center a welcoming and fun place to work, as did Ann Townes at the Olin Institute and Susan Lynch at the Belfer Center.

I have benefited from an extremely supportive group of colleagues since I arrived at the University of Pennsylvania in fall 2007, especially Avery Goldstein and Ed Mansfield, but many others as well. The Browne Center for International Politics at the University of Pennsylvania has generously supported the final stages of my research and writing process. An earlier version of chapter 6 appeared in *International Organization* 64, no. 1.

Chuck Myers has been a terrific editor at Princeton University Press, and his suggestions have improved my book tremendously, as have the edits of Cindy Milstein and work of Kathleen Cioffi. I would also like to thank my undergraduate adviser, Dan Reiter, who encouraged me to go to graduate school and become a professor.

My incredible wife, Rebekah, has read and commented on every word of this book, providing invaluable support along the way. Finally, I dedicate this book to my parents, Linda and Martin Horowitz. It would not have happened without their love.

The Diffusion of Military Power

Chapter 1

INTRODUCTION

INNOVATIONS in the production, deployment, and application of military power are crucial to international politics. Unfortunately, most assessments of the international security environment fail to incorporate either the relevance of military innovations or the importance of their spread. For example, in a thirty-year period, from 1850–80, the French Navy became the first to develop shell guns, and the first to deploy a steam-powered warship, an ironclad warship, a mechanically powered submarine, and a steel-hulled warship. These developments should have helped the French Navy gain superiority over its bitter rival, the British Royal Navy, but they did not. Moreover, barely a decade after the introduction of the steel-hulled warship in the 1870s, a new innovative school of naval theorists in the French Navy argued that the future of naval power lay with emerging technologies like torpedo boats and submarines, not the battleship. France was going to jump ahead once again. Yet despite this foresight and demonstrated initiative, most people generally do not consider France a great naval innovator of the period. Why is this? What advantages did it get from its introduction of a series of useful technologies into naval warfare?

The real answer is that the French Navy received no advantage. Unlike the U.S. Navy, whose mastery of the technology *and* organizational practices associated with carrier warfare provided it with a sustainable edge in naval power in the second half of the twentieth century, the French could not institutionalize their advantage. While the French excelled at inventing new technologies, crippling organizational debates prevented the integration of those technologies into French naval strategy. In each case, the French were the first to introduce a new naval warfare capability, while the British Admiralty appeared, in public, disinterested in French developments. Yet in each case the British, who had been carefully studying French advances in private, quickly adopted the new capabilities, improved on them, and used Great Britain's superior industrial production capabilities to eliminate France's ability to gain a relative power advantage from its inventions.

A prescient analysis in 1902 of submarine warfare by Herbert C. Fyfe, the "Sometime Librarian of the Royal Institution, London," includes an appendix on the French Navy that expresses French feelings on the matter:

> "We have seriously believed," says a writer in the *Journal de la Marine*,
> "that in all the great modifications that have been brought about in the

construction of submarines is the result of the important changes which the last fifty years of the century have produced in the art of naval warfare. All these changes have been sought out, experimented upon, studied, and finally realized by France, who has also been the first to apply them. These results have established in a brilliant and incontestable manner the skill of our engineers; but our rivals have not only appropriated the results of our labours, but they have not been slow to place themselves on equal terms with us, and finally to excel us in the application of these discoveries.... We have been only the humble artisans working for them to establish their superiority." (Fyfe 1902, 281)

While France was the technological first mover in several cases, it failed to harness its advances into an actual war-fighting innovation in a way that increased France's relative naval power.[1] Instead, it was the British Royal Navy that came to exemplify naval power in the mid- to late nineteenth century as it entered an era of naval superiority.

The failure of the nineteenth-century French Navy to exploit its technological inventions in sea power yields two important lessons for a general understanding of military power and international relations. First, inventing technologies or even being the first to use them does not guarantee advantages in international politics. There is a big difference between the introduction of a technology on to the battlefield and the full integration of that technology into national strategy, including warfare and coercive diplomacy. It is the difference between the two, in fact, that often determines success or failure in international politics. It is the employment of technologies by organizations, rather than the technologies themselves, that most often makes the difference.

Second, in contrast to most prior work on military innovation, which has tended to focus on who innovates and why, it is the diffusion of a military innovation throughout the international system that most determines its influence on international politics. The study of military power is incomplete at best without a theoretically coherent understanding of how states respond to major military innovations, and how the pattern of their responses helps drive the rise and decline of nations as well as the patterns of warfare frequently analyzed by other scholars. By developing a theoretical framework that can bring together empirical topics like suicide bombing and carrier warfare that scholars have tended to study separately, this book presents a new, more efficient, way to think about approaching the diffusion of military power.

The introduction and spread of new means of generating military power, sometimes called major military innovations (MMIs), have played a critical role throughout history in determining the global balance of power along with

---

[1] This introductory section is based on both Fyfe's and Theodore Ropp's work on the French Navy (Fyfe 1902; Ropp 1987, 8–11, 42).

the timing and intensity of wars.[2] The infamous Mongol armies, with their mastery of the composite bow and a new form of cavalry strikes, toppled nations from China to those in the eastern part of Europe because of their leaps in technology and strategy. Hundreds of years later, the German debut of blitzkrieg warfare at the outbreak of World War II helped them rout French forces and consolidate control over Western Europe. But despite their significance in terms of driving change in international politics, the processes that govern the spread of innovations and their effects are little understood in the field of international relations. Several questions about military power remain unanswered: Is it best to be the first mover, to borrow a term from economics, and the first to figure out how to effectively employ new types of military power, like the Germans with blitzkrieg?[3] Or is it better to be a follower, learning from the leader, and trying to extend and improve the original ideas, like the Germans with all-big-gun battleships responding to British innovations? How do nonstate actors fit into this story? Insurgent and terrorist groups have to make decisions about military strategy just like nation-states. How do they decide whether or not to adopt new innovations in how they use force like suicide bombing?

This book addresses the broad puzzle of why some military innovations spread and influence international politics while others do not, or do so in very different ways. These patterns are explained with a theory of the spread of military power called adoption-capacity theory.

Nation-states have a number of possible strategic choices in the face of military innovations. These include adoption, offsetting or countering, forming alliances, and shifting toward neutrality, as noted in the preface. Adoption-capacity theory posits that for any given innovation, it is the interaction of the resource mobilization challenges and organizational changes required to adopt the new innovation, and the capacity of states to absorb these demands, that explains both the system-level distribution of responses and the choices of individual states.

As the cost per unit of the technological components of a military innovation increases and fewer commercial applications exist, the level of *financial intensity* required to adopt the innovation increases. The rate of adoption decreases and alternatives like forming alliances become more attractive. Similarly, if an innovation involves large-scale organizational changes in recruitment, training, and war-fighting doctrine, the innovation requires a high level of *organizational capital* for adoption, and fewer actors are likely to adopt it. Some states will have the necessary capacity and interests, while politics will prevent adoption by others. If capacity and interest are lacking, no matter how

---

[2] Chapter 2 discusses defining and operationalizing military innovations.

[3] As chapter 2 describes, in the blitzkrieg case, while the British were the first movers with regard to the technology, the Germans debuted the mature innovation.

intrinsically compelling a new innovation may seem, it will not diffuse throughout the system. Accurately measuring these variations in diffusion also more effectively explains shifts in the balance of power and warfare than traditional theories alone can do. While higher financial requirements generally mean that the adoption patterns will benefit preexisting wealthy and powerful states, higher organizational change requirements can handicap the wealthiest states and upset the balance of power toward newer and more nimble actors.

The question of how states deal with periods of uncertainty about military power is of special interest today. Significant global economic turmoil now accompanies ongoing debates about the future of warfare in the information age. International relations scholars have demonstrated that uncertainty about the current and future security environment can be a primary cause of conflict (Fearon 1994a; Powell 1999; Smith and Stam 2004). Sharp debates exist between those who believe that the United States should optimize its military for future counterinsurgency campaigns like Afghanistan and Iraq, and those who believe that the United States should focus instead on its conventional capabilities (Gentile 2008; Mazarr 2008; Nagl 2009). An important wild card for both perspectives is the role of the information age in international conflict.

The information age is popularly described as the application of information technology to enhance the productivity of businesses and government, increasing the ability of societies to rapidly create and disseminate large amounts of information anywhere around the globe in real time. The information age, like the Industrial Revolution before it, will eventually have a large-scale impact on warfare.

While some degree of change is likely inevitable, the details of that change and the consequences are still very much in the air. In particular, the United States currently appears to lead the globe in developing and integrating information age advances into its military forces. But software-heavy developments may come to dominate the information age, rather than expensive physical hardware. The declining cost of computing technology, Internet access, and devices like personal GPS units, along with the dual-use nature of many information age military technologies like precision-guided munitions, mean new capabilities may become available to an increasing number of countries over time. While the United States has led the way in utilizing information technology in its military operations, its lead is far from assured. Peter Singer (2009) has described the way that the robotics revolution will impact the future of warfare, contending that there are risks for the United States as well as potential benefits.

In that hypothetical case, the U.S. government's devotion to its tanks, bombers, and carriers could become an albatross that drags down the U.S. military, which might face organizational challenges in transforming itself, in favor of states that figure out new and better ways to organize their forces to take advantage of information age technologies. Countries like China and India could end up leapfrogging a U.S. military that is increasingly focusing on irregular forms of warfare like those in Afghanistan and Iraq. Such an outcome is not on the immediate horizon and is far from inevitable, but it is a mistake to think that the United States is guaranteed to have the strongest conventional military forces in the world. These changes will also potentially empower nonstate actors attempting to find new ways to mobilize and fight against nation-states. Terrorist groups are already shifting the locus of their education, recruitment, and training operations to the "virtual" world of the Internet (Cronin 2006, 83–84; Hammes 2004, 198–99). The empowerment of nonstate actors means that a world of information warfare could substantially increase the capacity of terrorist groups and insurgents to deliver disruptive strikes on the major powers. Potential examples include taking down electricity grids or reprogramming satellites, which would further increase security challenges. While adoption-capacity theory cannot purport to provide exact answers, it can help us predict future trends and know the right questions to ask.

In sum, different military innovations spread throughout the international system differently, and the way they spread has a large effect on key issues in international politics like the balance of power and the probability, intensity, and length of wars. Understanding the spread of military power is therefore important not just for international relations theory but also for policy analysts interested in the future of global power and U.S. strategy as well.

Why the Spread of Military Power Matters

Military power is the measure of how states use organized violence on the battlefield or to coerce enemies. It represents the combination of the technology used to fight—"hardware" such as rifles, artillery, and bombers—and the organizational processes used to actually employ the hardware—"software" like recruiting and training. It is tempting, however, to view the spread of military power as simply the spread of military technology, the tools and devices used to prepare for or fight armed conflicts (Zarzecki 2002, 74).[4]

In contrast, in this book I am concerned with the spread and impact of changes in the character and conduct of warfare. While technological change often accompanies the innovations we remember in history, technology alone is rarely enough. Instead, building on work by Emily O. Goldman and others, it is the way militaries take raw technologies and use them that creates military force and influences diffusion patterns (Goldman and Eliason 2003a).

---

[4] This is also the starting point for most studies of arms races as well as arms imports and exports. The focus on quantitative measures of technology perhaps initially occurred simply because tanks and rifles are easier to count than methods of recruiting and training (Farrell 2005). For more on other theories of diffusion, see chapter 2.

My approach draws on evidence from the business world that shows studying technology alone is not enough to capture the essence of how innovations matter and what makes successful change more likely. For example, in the 1990s, Dell Computers pioneered a model of production that relied on made-to-order computers based on customer specifications, leading to lower inventories and overhead costs than its major competitors. This innovation in its organizational structure improved Dell's ability to integrate exogenous, or external, changes in personal computer technologies. When a technological change occurred, like the release of a new microprocessor from Intel, Dell could integrate it into its consumer production lines within a matter of days and without significant outdated warehouse stock; it generally took weeks for its competitors to do the same. This gave Dell an enormous advantage in its ability to deliver top-notch products to its customers, leading to more sales (Brynjolfsson and Hitt 2000, 29–30). While the technology mattered, since new microprocessors produced changes in computers in ways that altered costs and orders from customers, every computer company received the same chips from Intel. It was Dell's ability to integrate the new technology more efficiently than its competitors that produced its market advantage.

Another example of why both technological and organizational resources matter comes from survey data on business productivity. In 2001, the McKinsey Corporation and the London School of Economics surveyed over one hundred businesses that implemented technological changes, changes in managerial practices, or changes in both areas. The results showed the discontinuous impact of combining organizational and technological change. Businesses that implemented exclusively technological changes experienced a 2 percent increase in productivity, which paled in comparison to the 9 percent increase generated by exclusively managerial changes. Yet businesses that adopted both managerial and technological changes experienced 20 percent productivity increases, almost double the total from adding together technological and managerial change (Dorgan and Dowdy 2004, 13–15).[5] These results explain why companies like Dell succeeded in the 1990s and Apple has done so over the last decade.

Many international security researchers rely on measures of national power like iron and steel production, the numbers of troops or the defense budget of leading states, and their populations. The National Material Capabilities data gathered by the Correlates of War (COW) Project includes information on the military, industrial, and demographic capabilities of each state, which is summed into the Composite Index of National Capability (CINC).[6] CINC-based

[5] Despite many differences, military organizations share some key facets with firms, including the need to compete with other actors, the threat to survival from competitive failures, the development of bureaucracies to regulate and manage their operations, and the need to make strategic choices in response to changes in the external environment (Cronin and Crawford 1999; Waltz 1979).

[6] For more on CINC data and the COW Project, see Correlates of War 2 Project 2006; Singer 1987; Singer, Bremer, and Stuckey 1972.

research has become the standard way to measure power in international relations scholarship. The use of CINC data has produced a number of important insights into international politics, including evidence that materially stronger and wealthier states are more likely to win wars, all other things being equal, and that system power concentration is significantly related to militarized disputes (Bennett and Stam 2004).[7]

A growing body of literature in international relations, however, suggests that measuring military power and predicting military outcomes involve more than simply assessing the material resources states can bring to bear on the battlefield. Studies in recent years using more sophisticated quantitative models have built on some of the early research and shown that simpler models only relying on material power indicators do not reveal the full picture. For example, work by Dan Reiter and Allan Stam (2002) focuses on the political regimes of states, and how they influence battlefield outcomes.

Additionally, research by Stephen Biddle (2004, 21; 2007a, 218–20) demonstrates that material measures of international power are not in and of themselves enough to predict the outcomes of military campaigns. Biddle argues that force employment, or what militaries do with the equipment they have—the decisions they make about how to organize and deploy their resources—plays an important role in determining the military power of states (see also Stam 1996).

Materially strong states with weak force employment concepts sometimes lose, while materially weak states with strong force employment concepts sometimes win. For instance, despite having more ships, guns, and people, the Russian military lost badly to imperial Japan in the Russo-Japanese War. Another example of a sure loser according to conventional measures of military power is Israel, which confounded material indicators in a series of wars against numerically and materially superior Arab foes attacking from multiple sides. These cases show that core issues of international security cannot be explained without reference to much more than the number of people and specific technologies involved (Brooks 2007, 228).

Still, understanding the importance of both organizations and technology in producing military power is only the first step in appreciating the way military innovations influence international politics. The second part of the puzzle is the differences in the capacity of militaries to successfully adapt to changes produced by those innovations. New military innovations are not created equal when it comes to the ease of adoption. For example, the technological

---

[7] These systematic tests often navigated competing claims in the qualitative literature to help move intractable debates forward. The military dimension of both the Organski-Davies total output model and the Singer-Bremer-Stuckey national power model, for example, rely on military expenditures and personnel as the most important measures of military capabilities (Organski and Kugler 1980, 31; Singer, Bremer, and Stuckey 1972).

components of some innovations, like nuclear weapons, are extraordinarily expensive, especially for first movers and early adopters. In contrast, the unit costs of the technological components of some other innovations, like the rifle or machine gun, are relatively inexpensive. The organizational change requirements of innovations can also vary widely. Utilizing chemical weapons in World War I involved adding them into existing operational plans, not fundamentally changing the way militaries organized themselves. In contrast, adopting Napoleonic mass mobilization required an enormous shift in how militaries recruited and trained as well as the use of the division structure and the creation of skirmishers. It is these shifts in the financial and organizational requirements for adopting innovations—given the different capacities of military organizations—that produce varying implications for the international security environment (Gilpin 1981, 63).

Diffusion begins when a major military innovation reaches a critical "debut" or "demonstration" point. These terms, drawn from studies of business strategy, refer to the point when the relevant community has sufficient information to reasonably understand the significance of an innovation. While much of the time innovations debut through a demonstration during warfare, sometimes the revelation of a new capability during peacetime is enough to trigger a response, as when the British Navy introduced the *Dreadnought*.[8] This can vary depending on a variety of factors. The most critical of these is the extent and success of efforts by the first mover to shield knowledge of how the innovation works from potential adversaries or other states once it recognizes it has developed new military capabilities. Sometimes militaries do try to hide crucial elements of advances from the international community, as the Royal Navy did when it introduced the *Dreadnought* or the United States did with the Manhattan Project even after dropping two atomic bombs. At other times, an innovation debuts in a relatively transparent fashion, as it did when the United States and Japan both placed the aircraft carrier at the center of their fleets in the midst of World War II.[9] The debut point where diffusion starts varies from innovation to innovation.

Explaining the Spread of Military Power

Adoption-capacity theory combines research on the way both militaries and businesses change with new insights into the relative costs of new military

---

[8] The differences between introducing innovations in wartime or peacetime may be relevant as well.

[9] The question of shielding new military capabilities is related to Robert Axelrod's discussion (1979, 231–32) of when to debut new weapons in the first place. Whether or not the first mover can keep the innovation secret may be more or less possible depending on the domestic and international political environments.

systems to explain how military innovations spread once they have been introduced into the international system.[10] The basis of the theory is recognizing the adoption-capacity requirements of an innovation and how the capacities of individual states measure up.

Approaching the spread of military power from this perspective sidesteps the traditional debate about whether strategic competition, cultural factors, or norms best explains emulation and allows for the construction of a more powerful new theory. There are many reasons why states are interested in adopting innovations: strategic necessity, international norms, cultural openness, the need for interoperability with allies, and many others. Threats are a vital part of the matrix of factors that motivate nation-states. It is even possible that for the states that initiate military innovations, threats play an important role in their drive and capacity to innovate.[11] Prior research, though, has often presumed that for potential adopters, where there is a will to adopt, there will be a way (Elman 1999; Resende-Santos 2007). In the real world, states are sometimes overmatched no matter how well they optimize, and sometimes states do not adopt innovations even when they face large threats. Rather than viewing capabilities as totally fungible depending on state strategy, at least in the short- to medium-term it might be financial and organizational constraints that shape possible strategies as well as the probability of success.

Adoption-capacity theory argues that, once states have the necessary exposure to an innovation, the diffusion of military power is mostly governed by two factors: the level of financial intensity required to adopt a military innovation, and the amount of organizational capital required to adopt an innovation. As briefly introduced above, financial intensity refers to the investments required to purchase the physical hardware associated with an innovation, along with the relative ability of states to make those investments. Key to determining the level of financial intensity required for adoption is whether the relevant technology is exclusively military or has commercial applications, and the cost per unit of the physical hardware associated with the innovation, like a battleship or an aircraft carrier, in comparison with previous procurement. The more military oriented the technology and the higher the unit cost, the higher the financial intensity required for adoption.[12]

---

[10] The notion of relative costs here is complementary to measurements of military balances that incorporate the way new military advances can influence the relative cost of war (Anderton 1992; Powell 1999, 110–12, 197–98). James Fearon (1995a, 6–8) in particular explicitly recognizes the way that assumptions about rapid emulation have influenced offense-defense debates. Instead, in his view, what matters is the relative pacing of adoption.

[11] Barry Posen (1984, 1993) argues that threats drive the innovation process by determining whether civilian intervention occurs.

[12] This is related to research on capital intensity, but rather than focusing on the trade-offs between labor and capital, financial intensity is more about the way that capital is invested (Gartzke 2001).

The other half of the new theory is organizational capital, the intangible change assets needed by organizations to transform in the face of major military innovations. The study of organizations in general and military organizations in particular is hamstrung by the idiosyncrasies of individual militaries and the difficulties involved in parsing out exactly what determines their propensity to change. Though this will always be an issue, there are a variety of different ways to measure and evaluate the capacity of military organizations to change. Organizational capital is an imperfect but powerful way to conceptualize the potential change capacity of a military organization. Three factors in particular, measurable in military organizations prior to the demonstration of a given innovation, appear to best predict whether or not the organization will have the necessary capacity to adopt. First, the amount of resources devoted to experimentation is an indicator of the willingness and ability of organizations to consider major innovations. Second, as Mancur Olson (1982) contends in economics, older organizations often become bureaucratically ossified as subgroups of control proliferate, generating an increasing number of veto points that prevent innovations from being adopted. Therefore, the longer military organizations last without experiencing serious upheavals such as regime changes from within or defeats in interstate wars, the worse they should be at integrating innovations.[13]

Finally, the way that military organizations define their critical tasks plays a vital role in defining the range of the possible for those organizations (Wilson, 1989). The broader the definition of the organization and its purpose, the better it will be at adopting innovations. Al Qaeda's willingness to consider any and all operational methods for attacking the United States and its allies made it nimble enough to adopt suicide terrorism. Al Qaeda defined the means it would use to achieve its goals very broadly. In contrast, when an organization narrowly defines the optimal means to pursue its goals, the chances get higher that pushback from elites within the organization will prevent the adoption of innovations. A textbook case of how a limited view of the means to success can negatively influence an organization is the U.S. Army during the Vietnam War. The Vietnam era U.S. Army viewed using superior firepower as not just a means to an end but rather an end in itself; its ability to employ that firepower bled into how it measured success and failure. In 1965, the army even defined success based on the generation of enemy casualties.[14] This made it difficult for the army to master counterinsurgency operations requiring a lower emphasis on lethality (Krepinevich 1986, 5). The army instead preferred search-and-destroy

---

[13] There is no necessary correlation between organizational age and size. Nevertheless, older organizations are more likely to produce special types of bloated bureaucratic structures that make change difficult.

[14] It is even possible, based on Scott Gartner's work (1997), to argue that the focus on overwhelming firepower might have influenced the army's choice of body counts as its dominant indicator, or metric for success, during the Vietnam War.

missions where it could apply maximum firepower and generate the largest number of casualties (Gartner 1997, 130–31).

The speed and extent of an innovation's spread therefore depends on the relative financial and organizational requirements. Those requiring less to adopt will spread faster than those that require more. Adoption-capacity theory shows, however, that the levels of financial intensity and organizational capital required to adopt an innovation not only significantly influence the rate and extent of its spread throughout the international system but also drive its affect on international politics. Since it is generally easier to adopt the physical technologies associated with an innovation than the overall system of fighting, innovations featuring especially high levels of financial intensity are likely to spread, albeit slowly. In particular, financially intense innovations requiring organizational changes that sustain, rather than disrupt, previous critical tasks are likely to spread gradually but consistently, benefiting the preexisting strongest and wealthiest states in the international system. While preinnovation major powers lacking the financial capacity to adopt are likely to slip and become second-rate powers, the innovation is unlikely to reorient systemwide power balances. In contrast, innovations requiring disruptive organizational transformations but relatively reasonable financial investments, like blitzkrieg, the German combination of the radio, airplane, tank, and other motorized vehicles, will spread haltingly, with only a few states adopting the full innovation, and most acquiring some of its technical components but not adopting the new system of warfare. Innovations requiring large degrees of disruptive organizational change most clearly create strategic openings for power transitions and generate larger first-mover advantages. New powers that master the necessary organizational changes can gain an advantage over their potentially bigger though less nimble major power opponents.

Essentially, new major military innovations can create discontinuities in international politics, ushering in the risky situations described by Robert Powell (1999, 85, 199) where the actual balance of power sharply diverges from the distribution of benefits in the international system, because the system has not yet caught up to the new power realities. If a rising power develops a new innovation, it may gain an enormous edge in its drive to the top. In response, status quo powers that can quickly mimic and adapt to new military innovations or respond with their own new innovations have the best chance of limiting the disruptive impact of the innovation as well as maintaining their relative power level in the face of a challenge (Gilpin 1981, 60–61, 161–62). Sometimes, however, new major military innovations confront major powers, but for financial or organizational reasons they cannot adopt in the short- to medium-term. This presents a major power with a fundamental choice: continue posturing as if it is a major power, or recognize the writing on the wall and seek an alternative strategy that may involve making its interests subsidiary to those of another likely adopter. When states choose the former path, like the Austro-Hungarian

Empire did before World War I, it can destabilize the international system by accentuating informational gaps in national analyses of likely war outcomes. The resulting gap between beliefs and reality are a common cause of war because they make miscalculation and escalation more likely on all sides.[15]

Adoption-capacity theory is also useful for explaining the behavior of non-state actors, as chapter 6 highlights. Like conventional militaries, insurgent and terrorist groups must make decisions about resource allocations and the organization of their forces. Financial intensity and organizational capital are useful metrics for understanding the strategic choices of terrorist groups in the suicide terrorism era. Those groups with critical tasks based in particular operational methods and that existed long before the beginning of the modern age of suicide terror faced substantial hurdles to adopting the innovation. The PIRA and ETA never adopted, for example, while it took Fatah, a key part of the Palestinian Liberation Organization (PLO), nearly twenty years to adopt. In contrast, groups with younger organizational ages and less defined critical tasks, from the broad tactical setup of the Tamil Tigers to the cell-based network of Al Qaeda, were more easily able to take advantage of suicide tactics, providing them with a new weapon.

Just as the tacit knowledge required to effectively operate aircraft carriers creates significant organizational obstacles for countries interested in adopting carrier warfare, the availability of instruction in suicide methods or direct geographic proximity to suicide terrorists has constituted a tacit barrier to entry for some terrorist groups. This is not to say that variables like national liberation movements, religion, and/or fighting for popular influence are not motivating factors for the adoption of suicide terrorism. Rather, adoption-capacity theory can explain both the groups that have adopted and why other groups do *not* adopt suicide terrorism—something prior work has rarely addressed. Prior theorizing on terrorist strategy, like that on military innovation, has tended to focus on what drives the interest of terrorist groups in suicide bombing, implicitly assuming that the desire to adopt suicide terrorism is enough to make it happen (Pape 2005). Applying adoption-capacity theory to the case of suicide bombing shows the web of interconnections between groups and the flaws in trying to predict terrorist group behavior without an understanding of the broader linkages between groups.

It is important not to overstate the scope of the theory. There are a variety of reasons why states are interested in innovations, why states adopt innovations, and why states become more or less powerful. Hopefully this book can make a contribution to ongoing debates in the academic and policy worlds about what types of changes are more or less likely to occur in periods of uncertainty about military power.

---

[15] This argument builds from work on bargaining, information, and war. In particular, see Fearon 1995b.

Importance for the Future of Warfare

While this book is mostly focused on the military innovations of the past, looking forward to the future is useful both as a demonstration of the theory and to show its relevance to ongoing policy debates. Adoption-capacity theory can help explain the way different types of warfare in the future will provoke different types of reactions on the part of responding actors, and benefit or disadvantage different states. This can help provide a framework for discussion by showing how the likely implications for the security environment depend on particular assumptions about the future.

At present, there is a spirited debate occurring around the world about the types of wars most likely in the future. In the United States, the debate about the utility of "network-centric warfare" has given way to one about whether the United States should focus its limited resources on institutionalizing the hard-earned counterinsurgency lessons of Afghanistan and Iraq, or refocus the U.S. military back toward conventional warfare (Boot 2006, 8–9; McMaster 2008, 25–28). Biddle maintains that there is not enough evidence to overturn the prominence of the "modern system" of warfare—the use of firepower, cover, and concealment to take and hold territory in conventional land engagements. Focusing on the dangers of allowing technology to dictate force structure, he states that the modern system is the "orthodox" approach to war, and that scholars and policymakers should be careful before embracing "heterodox" approaches (Biddle 2007b, 463–64). He also notes that many instances viewed as counterinsurgency campaigns, like Lebanon, have actually revolved around the application of conventional modern system principles (Biddle and Friedman 2008).

Thinking that the information age will make a difference in future warfare does not mean excluding the human element or skill on the battlefield. Nor should believing in the human element along with the importance of tactical proficiency or skill mean ignoring the way that the information revolution may shape the realm of the possible in warfare (Gray 2006). One popular and persuasive perspective, promoted most clearly by Frank Hoffman (2007), argues that the future of warfare will be "hybrid," demonstrating facets of both regular and irregular wars, but in an operating environment characterized by the information age.

If the information age, like the Industrial Revolution before it, is likely to have wide-reaching and complicated effects on society, determining its impact on the security environment matters whether the most probable future combat scenarios are potential U.S.-China scenarios in East Asia, land-heavy wars in the Middle East, or quasi–peace enforcement operations around the globe. It is possible that the most likely wars of the future are irregular campaigns featuring land forces, but that there are also significant possible contingencies involving the heavy use of naval and air forces. The impact of the information age on each of these might be different, just as it might be different for states and nonstate actors.[16]

The U.S. military struggled during the early part of the last decade learning how to fight against insurgents in Afghanistan and Iraq. Yet many who think about the future of military power argue that at the very least, the U.S. military will indefinitely maintain and deepen its conventional military superiority (Brooks and Wohlforth 2008, 27–35; O'Hanlon 2000, 168–69). Acclamatory statements that the U.S. military has already mastered the information age are surely overstated; the United States leads the world in the application of information technology to its military operations, but U.S. military operations over the last decade suggest that the United States has far from mastered the information age. There are several areas where disruptive changes could influence the trajectory of warfare.

At present, the U.S. military has made great strides in precision warfare— the use of new communications and guidance technologies to hit targets more accurately as well as at greater distances than ever before. These capabilities are cited by both counterinsurgency and conventional war advocates as important for the future, although there are disagreements about their relative effectiveness at present. Whether or not these advances are properly categorized as an MMI is a matter of debate. But the initial demonstration of U.S. precision-guided munitions in the Gulf War may have represented the starting point of the ticking innovation clock, like the debut of aircraft carriers by the Royal Navy and/or the introduction of the tank by the Royal Army in World War I, rather than representing a completed, fully functional MMI (Welch 1999, 122). Linear advances in precision warfare up to this point have extended the edge of the U.S. military at conventional operations. Utilizing the innovation requires expensive platforms like bombers and ships, meaning there is a high level of financial intensity required to adopt. Conducting counterinsurgency operations has proven challenging for the U.S. military, however, due to the large-scale organizational challenge of shifting the armed services away from focusing on overwhelming firepower.

Precision warfare, under most foreseeable circumstances, will initially remain costly even as the reliance on major weapons platforms like bombers decreases. The financial intensity required to implement anything like what the U.S. military does today is so high that even a small decrease in unit costs will not allow many more states to actively seek military dominance. Moreover, as long as the

---

[16] While critics of network-centric warfare, like Frederick Kagan (2006, 389–90), and scholars studying warfare in the information age, such as Peter Dombrowski and Eugene Gholz (2006, 4–6), are certainly right that the information revolution will not necessarily lead to one particular optimal force structure outcome, that doesn't mean the changes wrought by the information age are irrelevant or that the information age is unlikely to matter at all.

core platforms for using precision warfare are linked to the platforms of today, recruiting, training, and organizing modern militaries will look similar.

Advances in areas like robotics and information technologies such as computing could shift the military power status quo, though. If advances in munitions and especially unmanned vehicles begin to make the expensive launching platforms that sit at the core of the U.S. military irrelevant, it could risk large-scale changes in military power balances. If a cargo ship or cargo plane is suddenly just as good for launching a missile at a target as an F-15, the financial intensity requirements for implementation will drop and the organizational capital requirements will increase.[17] In such a situation, militaries may also have to recruit differently—recruiting unmanned aerial vehicle (UAV) pilots who excel at video games instead of "fighter jocks," for example—and train people to conduct different tasks, since they will be operating mostly with joysticks rather than in actual battle spaces with the enemy.

If rising U.S. military capabilities illustrate a path away from the financially intense platforms that currently help ensure U.S. dominance, while also requiring militaries to organize themselves differently to best take advantage of available capabilities, countries such as China and India could find it increasingly possible and attractive to militarily compete with the United States. This low level of financial intensity and high level of organizational capital required to take advantage of the information age in the area of conventional war would then lower the barriers to entry for potential competitors to the U.S. military. Other states in the international system could then acquire the necessary capabilities to begin effectively reducing the military edge of the United States unless the country continues innovating to stay ahead. Possible areas for development include not only robotics, with UAVs as the most obvious manifestation, but cyberwarfare as well. The resulting situation could cause major shifts in the global balance of power as some states benefit and others, unable to implement, are increasingly left behind. Adoption-capacity theory principles help explain the microfoundations underlying the concern by authors like Max Boot (2006) and Singer (2009) about the way the information age could boomerang, allowing other countries to catch up to the United States in the long run.[18]

Just as it will influence warfare between states, the information age is likely to influence the trajectory of actions by nonstate actors. The commercial spread

---

[17] This could occur because of increases in range and the miniaturization of tasks currently conducted by support planes like the Airborne Early Warning and Control System into next-generation missiles.

[18] This is true whether one wishes to call this sort of development the second stage of precision warfare or something as yet undetermined but related to the combination of materials, information technology, and communications (Kagan 2006, 395).

of Internet access around the world along with the low unit cost of basic computers and laptops mean that any nonstate group with a minimal level of financial support can establish a Web presence that is useful for coordination, communications, and planning (Arquilla and Ronfeldt 2001).

Given that information age innovations are likely to feature low required levels of financial intensity for adoption, it may open the door to the acquisition of key components by nonstate actors as well. Cheaper and more widely available information age technologies could lower the barriers for groups seeking to challenge state authority, meaning it will become increasingly easy for new groups to spring up in virtual environments and to exchange information across borders (Hammes 2004, 207–9, 218). A proliferation of potential target points could foreshadow more dangerous cyberattacks against everything from the control systems at a power plant to the Department of Defense mainframe to Google. Groups will probably form faster, conduct operations, and potentially disappear, only to pop up again in another guise in another "virtual" place—or even another real place.

Again, these predictions are tentative. The information age may end up mattering quite a bit for some types of warfare, but much less for others. The point is that the debate should be about exploring the multiplicity of ways that periods like the information age may shape many different dimensions of warfare. Adoption-capacity theory is a useful tool to help explain the different outcomes likely in different security environments.

Moving Forward

Chapter 2, which follows, lays out what "counts" as a major military innovation and the theory of diffusion briefly explained above. This theoretical argument concludes with a discussion of the cases selected for analysis: British naval innovations in the late nineteenth and early twentieth century, carrier warfare, the advent of nuclear weapons, and suicide bombing. Chapters on each case follow the theoretical chapter. Two of the chapters—the ones on nuclear weapons (chapter 4) and suicide terrorism (chapter 6)—feature quantitative tests, while two—ones on carrier (chapter 3) and battlefleet (chapter 5) warfare—include more qualitative analysis featuring the use of both primary and secondary sources. Each empirical chapter concludes with an examination of the way that the given major military innovation under examination influenced the international security environment, focusing in particular on power balances, the probability/duration of wars, and alliance patterns. The importance of the variables identified by the new theory are compared with explanatory mechanisms from alternatives described in chapter 2.

I test adoption-capacity theory through a multimethod approach to make the overall results more reliable. Additionally, by using rigorous social scientific methods to study a topic of substantive interest to both academics and policy-makers, this study attempts to cross disciplinary lines, and integrate theory and practice (George 1993; Goldman and Eliason 2003b, 22–23).

The conclusion (chapter 7) discusses the implications for scholarly analyses of international relations, and then evaluates claims about the onset of the information age and the ways it may influence the future of warfare. The conclusion describes some crucial issues that are often absent in debates about the future of American defense strategy, and how adoption-capacity theory suggests that the information age may portend a much greater level of risk for U.S. conventional military superiority than some previous authors have envisioned.

Chapter 2

A THEORY OF THE DIFFUSION
OF MILITARY POWER

WAR is a harsh teacher, Thucydides tells us. We either learn from others who are better at fighting than we are or we die. Yet there are puzzles. France knew after World War I that Germany remained economically and demographically stronger than France. Knowledge of the emergence of blitzkrieg warfare in Germany was available. In the 1930s, however, the French Army planned for the same kind of slow, methodical, defensive war that World War I had taught, even though the logic of its alliance system called for an offensive against Germany in the event of a German attack on Poland. Why? Egypt suffered a massive defeat in 1956 at the hands of the Israelis. It received massive financial and military technological assistance from the Soviet Union. It had every incentive to beat the Israelis at their own game, and the material means to do so. But it suffered a massive defeat again in 1967, for exactly the same reasons. France and Egypt should have learned from those countries that most threatened them, yet they did not. Why?

Scholars interested in military power have devoted serious attention to the question of how states try to gain advantages over other states through the creation of new ways of generating military power, also called military innovations. Barry Posen (1984) and Stephen P. Rosen (1991) have theorized about the conditions in which militaries are most likely to innovate, disagreeing about whether innovations happen as the result of pressure from actors outside military organizations, military mavericks within the system, or changing promotion patterns that empower those with innovative ideas.[1] This prior theorizing about military innovations in international politics, though, relies on clear implicit ideas about the way military power diffuses. By making those ideas explicit and engaging in empirical testing to determine which of them more accurately depicts world politics, this book explains how change in the international security environment occurs, advancing our understanding of military innovations beyond the first stage of the innovation life cycle.

Since we care about military innovations because of their consequences for international politics, it is necessary to focus on what happens *after* the initial innovation is demonstrated, not just on the innovation process. For example, blitzkrieg is more than an interesting historical footnote precisely because

---

[1] For a recent review article, see Grissom 2006. For examples of some key works on military innovation, see Avant 1994; Evangelista 1988; Kier 1997; Mahnken 2002; Pierce 2004; Posen 1984; Rosen 1991; Sapolsky 1972; Zisk 1993.

Germany used it to invade France and France proved unable to effectively respond. It is in the way states respond to innovations, the mechanisms governing the diffusion of military power, that the importance, or lack thereof, of innovations for international politics emerges.

DIFFUSION AND MILITARY INNOVATION

What Is Diffusion?

The spread of military power works by means of a phenomenon known in many fields as diffusion. The defining text on diffusion research in the social sciences, *Diffusion of Innovations* by Everett Rogers (2003, 11), describes diffusion as "the process by which (1) an innovation (2) is communicated through certain channels (3) over time (4) among the members of a social system." Critical to diffusion research is the pattern by which successful innovations generally spread throughout a population. Perceptions that an innovation is effective create the framework within which diffusion does or does not occur, meaning there is inevitably some linkage between innovation and diffusion. Normal innovations often diffuse in the shape of an S curve—an insight applicable across a wide variety of disciplines and areas, including agriculture, consumer products, public policy adoption, and military technology.[2] There are three essential stages of diffusion for "normal" products after a triggering event or debut. In the first stage, diffusion is slow as risk-acceptant actors, or early adopters, implement the innovation. In the second stage, once diffusion has reached a critical mass (analogous to a tipping point in game theory literature), the rate of diffusion speeds up and most of the set of actors with the ability to implement the innovation will do so. These are known as main adopters. In the third stage, laggards, or later adopters, implement the innovation. The total number of adopters, or percentage of adopters out of the pool of potential adopters, graphed over time, is known as a cumulative adopter distribution.

Innovations also often mutate as they spread and adopters alter them to the contours of their particular situation. For example, Geoffrey Herrera and Thomas Mahnken (2003, 242) show how, in the pattern of responses to Napoleonic warfare and Prussian open-order tactics in the nineteenth century, countries tended to adopt some key components of the innovations but not others, and modify the innovation depending on their requirements.

Studies of when the diffusion of ideas is more or less likely have also grown in recent decades, but the notion that there are interactions between domestic organizations and developments abroad is well established in the social sciences.

---

[2] Paul David's work on diffusion in the field of economics has also been significant. See, for example, David 1986.

Scholars ranging from Alexander Gerschenkron (1962), who wrote about competition and industrial development, to Peter Gourevitch (1978), with his argument about the way international politics conditions domestic economic decisions, have demonstrated the important connections between new practices abroad and decisions made at home. This research also shows the significance of developing theories that take into account the possibility that not all diffusion is efficient: some individual actors may not benefit from adoption, and some innovations are not relative improvements.[3]

Research in political science and related fields on diffusion has included work on domestic policy innovations (Berry and Berry 1990; Walker 1969), social policies (Mintrom and Vergari 1998), tort reform (Canon and Baum 1981), and tax policy (Berry and Berry 1992). Internationally, Beth Simmons and Zachary Elkins (2004) have studied the diffusion of financial policy, while others have looked at network effects in post-Communist states (Kopstein and Reilly 2000), Latin American social sector reform (Weyland 2004), international environmental regulations (Busch and Jörgens 2005), elections (Blais and Massicotte 1997), the impact of foreign direct investment on the spread of technology (Borensztein, De Gregorio, and Lee 1998), market-based infrastructure reforms (Henisz, Zelner, and Guillén 2005), gender mainstreaming (True and Mintrom 2001), and the spread of democracy (Cederman and Gleditsch 2004; Ward and Gleditsch 1998), among other topics.

Within the area of international conflict, some scholars have modeled the spread of warfare as a contagion process, drawing on theories from epidemiology (Simowitz 1998; Siverson and Starr 1991; Starr and Siverson 1998). Other studies of war and diffusion have included region-specific studies of Africa and Latin America as well as more general studies of the spread of coercive practices in international disputes (Bremer 1982; Hammarstrom 1994; Starr and Most 1983).

Studying the Diffusion of Military Power

Much of the existing research on the spread of military power comes from those who contend that strategic competition drives the diffusion process, states act primarily to maximize their own external security, and the most important determinant of behavior is the international environment. Applying his general insights on isomorphism, or convergence over time, to military technology, Kenneth Waltz (1979, 127) writes, "Competition produces a tendency toward

---

[3] David Strang and others have written about a wide variety of topics ranging from the spread of decolonization movements to the spread of poison pills in the corporate world to the student shantytown movement in the campaign for divestment from South Africa (Soule 1999; Strang 1991; Strang and Meyer 1993; Strang and Soule 1998).

the sameness of the competitors . . . [a]nd so the weapons of major contenders, and even their strategies, begin to look the same all over the world." Posen's work (1993) on the Prussian response to mass mobilization and João Resende-Santos's study (2007) of South American militaries of the late nineteenth and early twentieth centuries similarly show the way security challenges drive interest in adopting innovations.[4] Colin Elman (1999, 55–57) maintains that while security maximization can take the form of much more than just adopting an innovation, meaning Waltz's predicted isomorphism is unlikely to occur, the security environment plays a primary role in influencing state choices.[5]

Others evaluate the way that norms derived from "world culture," rather than competitive pressure, influence how states emulate (Farrell 2005, 451–52) or argue that the cultural tolerance of foreign ideas in general predicts the successful adoption of innovations (Goldman 2006, 70; Pollack 2002).[6] Research on the relationship between domestic politics, organizational factors, and military innovation has also played a central role in describing how militaries make decisions (Avant 1994; Evangelista 1988; Goldman and Eliason 2003a; Mahnken 2002; Zisk 1993).

This book attempts to use prior theorizing as a building block to explain how, given the distribution of geopolitical interests and national attributes, the financial and organizational requirements for adopting an innovation determine how it spreads. Prior research has generally addressed one piece of the diffusion puzzle at a time. In contrast, I take the existing literature as a jumping-off point to build a more cohesive theory that describes how innovations diffuse, how states respond to innovations, and how those patterns impact international politics. The question that mostly occupies the previous literature is what makes countries interested in responding to innovations, while the key question examined in this book is what determines the way these innovations end up spreading in general and actually influence the international security environment.[7] Adoption-capacity theory makes clear the distinction between interest in responding to an innovation and the substance of that response.[8]

---

[4] While some argue that studying emulation differs from studying diffusion because diffusion is descriptive while emulation is about the learning process (Resende-Santos 2007), in reality studying diffusion means evaluating the process by which an idea spreads, and emulation is one way that can occur—as a subset of diffusion.

[5] For an assessment of how strategic competition can influence the innovation process, as opposed to the diffusion process, see Goldman 2007.

[6] Theo Farrell's incredibly detailed work (2005, 470) on the Irish Army and its doctrinal choices after independence is an excellent example of the way that norms can sometimes influence the content of military strategy.

[7] For an exception, see Goldman's work (2003) on the development of carrier warfare, which recognizes the importance of capacity.

[8] The distinction between attempting to implement an innovation and actually doing so has been pointed out in the context of the difference between civilians ordering military reforms and militaries actually implementing them (Elman 1999, 39; Zisk 1993, 5).

Second, instead of focusing solely on whether or not states adopt innovations, adoption-capacity theory makes predictions about all the possible responses for a given state, only one of which is adoption. It also shows the limits to the assumption made by scholars like Resende-Santos about the preferability of emulation. Third, adoption-capacity theory uses the same theoretical premises to make predictions about how states will respond to innovations, the system-level distribution of responses, and the effect on international politics. Linking together how states respond to innovations and the consequences of those responses allows for broader insights into international politics. I return to the prior literature again below to look at the distinction between adoption-capacity theory and previous research.

MILITARY POWER AND MAJOR MILITARY INNOVATIONS

Defining Major Military Innovations

When the production of military power changes, meaning the character and conduct of warfare change in some measurable way, it is called a military innovation; the bigger the change, the bigger the innovation. MMIs are defined for the purposes of this book as major changes in the conduct of warfare, relevant to leading military organizations, designed to increase the efficiency with which capabilities are converted to power.[9] They are sometimes, but not always, closely related to changes in the technology used by military organizations. This book is concerned with new ways that military organizations generate power—and respond to innovations in the production of military power—not simply technological inventions.[10]

Given the possibility for misunderstandings about terminology that seem endemic to discussions of military innovations, some clarification is necessary. The "conduct of warfare, relevant to leading military organizations" refers to the way that leading states, sometimes called major or great powers, organize their militaries and plan to fight wars. The notion of changing the character of warfare refers to shifts in the core competencies of military organizations, or

---

[9] Where scholars most often disagree is about which shifts in military power should count as innovations. The definitional distinction between innovations being designed to help states get ahead and actually helping states get ahead is crucial (Gray 2002; Krepinevich 1994).

[10] This move is similar to the one made by Robert Gilpin (1981, 60). Approaching the question of shifts in power from the perspective of the diffusion of innovations also moves beyond the offense-defense balance paradigm, which makes sense since it is the relative balance of forces and beliefs about battles that influences behavior in militarized situations, not something inherent about particular technologies. Also, the offense-defense balance literature underplays the role of politics in determining behavior (Lieber 2005).

shifts in the tasks that the average soldiers perform.[11] Including "designed to increase the efficiency with which capabilities are converted to power" in the definition expresses the idea that all innovations are not the same and all are not successful. While the goal of major military innovations is to produce increases in military capabilities, whether that occurs is independent of the definition. This distinction is necessary to avoid tautology in the definition and measurement of the object of interest, the military innovation, and dependent variables like power balances and war.

How Major Military Innovations Emerge

Once we have determined what a military innovation is, we can move on to analyzing the way they debut in the international system and subsequently diffuse. Studies of the diffusion of consumer durables in the business world show that an incubation period often occurs between when new products enter the commercial market and when the product matures and sales take off. For example, the data on the spread of color televisions and compact disc players shows a large gap between the initial development of each product and when sales increased in the broader population (Golder and Tellis 1997, 257; Mahajan, Muller, and Bass 1990, 3).[12] The idea of an incubation period also makes sense in the military context, helping to distinguish between technological invention and a military innovation, because the specific technical capabilities that represent the public face of a military innovation are often introduced years, if not decades, before militaries develop them further and figure out how to use them to produce military power differently. In the case of the tank, for example, the technology was first used by the British at the Battle of Cambrai in 1917, but the mature innovation, which included the tank as well as a series of other technologies like radios and air power, did not debut until the use of blitzkrieg by the German military in World War II (Welch 1999, 122). In the intervening incubation period, the Germans were able to determine how best to utilize the new technology and reform their operational concepts to implement combined arms warfare.[13]

---

[11] This does not exclude the possibility that other actors might innovate. It is just to point out that an innovation reaches critical mass in a social system when key actors are forced to adapt in one way or another. The innovation well could come from below.

[12] Rajeev Kohli, Donald Lehmann, and Jae Pae (1999, 138, 141–42) speculate that the incubation period may function differently for discontinuous versus continuous product innovations.

[13] Prominent historian Ernest May finds Germany's success particularly impressive in comparison with France's failure. He describes France's military organization as "sclerotic" and unable to take advantage of an edge in the numbers of troops and tanks on the western front in 1940 due to massive intelligence failures by the Allies (May 2000, 10, 451–52). May's work also shows that simple measurements of forces on the battlefield like troop counts or the counts of particular technologies

Separating the incubation period from the takeoff period of a mature innovation reveals the importance of determining the *demonstration point* for
each major military innovation. The demonstration point for each MMI occurs
when the potential of its full capabilities is reasonably known in the international system through an action by a first mover, rather than the capability
merely being the subject of internal exercises or debates.[14] This is when the
clock begins ticking on potential adoption by other states.[15] An MMI exists
as a discontinuous change based in part on the perspective of the states that
view the demonstration of the innovation, not necessarily as a discontinuous
change for the first-moving state.[16] For example, while many German officers
viewed blitzkrieg as the logical culmination of German military advances in
the 1920s and 1930s, most French and British military officers saw it as a
radical departure from previous patterns of warfare on its debut (Mahnken
2003, 253).

Each innovation has a unique life span; different innovations have relevance
in international politics for different time periods. New developments, potentially even in response to a major military innovation, could trigger a new
MMI in a short period of time, constraining the time period of relevance for
an innovation. For instance, while the principles of warfare in the age of sail
stayed relatively constant for hundreds of years, the mechanized battlefleet system that replaced the sailing navy in the late nineteenth and early twentieth
centuries was itself replaced just a few decades later by carrier warfare at the
onset of World War II. Nuclear weapons have maintained their hold as a critical element of global power since 1945, a period of over sixty years, while blitzkrieg warfare supplanted trench warfare after only a few decades.[17] Given that
innovations can matter for varying period of times and different areas of warfare such as land or sea, it makes sense to study the innovation as the unit of

---

like tanks or planes do not always predict battlefield success. While the Germans had 76 first-line
divisions and 26 second-line divisions in May 1940, at the outset of the war, the Allies had 93 first-
line divisions and 30 second-line ones. Moreover, while the Germans had 2,439 tanks deployed on
May 10, 1940, the Allies had 3,079 (476–77). Instead, to supplement those measures it is also important to understand the way that organizations employ force—a conclusion this book draws as
well. The point here is not that quantitative measures are not useful, just that more sophisticated
models are sometimes necessary, and it is crucial to incorporate how militaries use raw materials.

[14] Many innovations debut in wartime. It is possible for an innovation to debut through the announcement of a new capability, revelations about doctrine or new technologies, or in other ways.
Warfare, however, is the most likely demonstration point, because states may save new innovations
for wartime and because the rest of the international community is more likely to be paying close
attention (Axelrod 1979).

[15] This distinguishes between diffusion processes and simultaneous developments by multiple
states. It is theoretically possible that more than one state could introduce the same innovation,
meaning there is more than one first mover.

[16] For more on differing perceptions, see Jervis 1976.

[17] For an alternative perspective, see Biddle 2004.

analysis, rather than trying to build something like a comprehensive data set that crosses several innovations.

Outlining Possible State Responses to Major Military Innovations

Once a military innovation debuts, those states interested in responding have a variety of options to choose from. Presume for a moment that states will generally respond to innovations in ways that they believe will maximize their foreign policy interests, however defined. The decisions of the state are influenced by a series of incentives and constraints that will shape its eventual response strategy. The geostrategic environment is one of the most significant factors determining the range of states interested in an innovation.[18] Other factors, ranging from international norms to cultural openness to the need for interoperability with allies and many others, may also influence interest in responding to an innovation.

Yet it makes sense to separate whether or not a state needs to respond to an innovation from the content of that response. That states may attempt to maximize their security does not necessarily suggest that they will adopt an innovation. Instead, as Elman (1999, 55–57) argues, while maximization could take the form of adoption, it could also take the form of relying on alliances and/or trying to counter an innovation through alternative means. While the question that some realists and others have traditionally asked about emulation may effectively answer why states are interested in responding to innovations, existing research has difficulty explaining what predicts the content of responses to innovations. Writing on the spread of military innovations or military technology traditionally views the strategic choices of states in response to innovation as a yes or no question—either innovations are adopted or they are not.[19] In reality states can pursue a much broader range of options, sometimes simultaneously, and those options may be preferable to emulation/adoption in some circumstances.

Implicit assumptions about the fungibility of capabilities have also driven the threat-based story about emulation—the idea that any state can adopt an innovation if it faces a large enough threat. But if capabilities were entirely fungible, it would make something like military victory logically impossible if both states optimized their militaries, since each state, facing a threat to its survival, would continually innovate and parry the thrusts of its opponent. The real world strongly suggests otherwise. To give an extreme case, no matter how a country like Belgium optimized prior to the Second World War, it could not defeat Germany. Especially in the short- and medium-term, we can

---

[18] See the work by Elman, Posen, and Resende-Santos cited above.
[19] See, for example, Resende-Santos 2007.

reverse the traditional story and assert that capabilities form a constraint that influences national preferences by defining the range of the possible. While the "rational shopper" approach assumes that, essentially, "where there is a will there is a way," the reality is more complicated.[20] If capabilities are not entirely fungible and the optimal response to an innovation is not necessarily adoption on the part of a threatened state, something more is required to unpack the diffusion process.[21]

Research on domestic politics and military organizations suggests that, like the strategic environment, domestic political and military organizational factors are vital in shaping the way military organizations behave.[22] Posen's work (1984) on military innovation, for example, indicates that both the external security environment and organizational factors can play important roles. Figure 2.1 highlights the range of potential responses that states have when thinking about how to deal with a new military innovation.[23]

One set of possible responses involves external actions, or changes that a state can make to its foreign policy to deal with the potential repercussions of another state adopting a military innovation.[24] A state could decide that the implications of the innovation make the achievement of national foreign policy goals no longer possible, requiring a lessening of overall foreign policy assertiveness or even a shift toward neutrality. This could lead to what Paul Schroeder (1994, 117) describes as "hiding" in international politics or potentially "transcending": pushing for international institutions or other means to deal with a situation in which a state no longer has the relative power to protect itself.

Another possible external response is to attempt to mitigate the costs of nonadoption by allying with a likely adopter (Walt 1987).[25] One option is balancing, creating or joining an alliance against the state that initially debuts the military innovation with another state or a group of states likely to successfully adopt or counter the innovation. The converse of balancing is bandwagoning, allying with the pioneering state that generated the innovation. Alliances boost the relative power of the involved states, because these states can work together

[20] "Rational shopper" is taken from Goldman and Ross (2003, 380).

[21] John Nagl's study (2005, 219) of U.S. and British counterinsurgency strategy in Malaya and Vietnam comes to a similar conclusion, arguing that the optimization of U.S. forces for conventional war after World War II made it difficult to adapt to counterinsurgency requirements in Vietnam.

[22] See, for example, the work by Evangelista, Zisk, and Mahnken cited above.

[23] This figure is not meant to suggest the paths are all exclusive, just to show the different possibilities. For a realist take, see Resende-Santos 2007, 69.

[24] This rubric assumes that a state has some time to adapt. The results might be different if a state faces a severe, short-term threat.

[25] This is similar to Powell's argument (1999, 176–79) that states sometimes may turn to alliances as a response strategy when potential adversaries experience increases in military capabilities. On how states make alliance choices, see also Snyder and Christensen 1990.

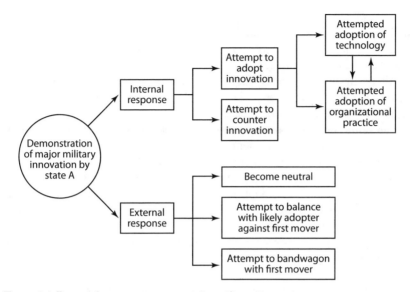

Figure 2.1. Potential state responses to major military innovations.

to do more than they could do alone.[26] Alliances may also help a state more rapidly gain access to the knowledge necessary to implement the innovation, either through buying time to build the capacity to adopt or through direct assistance from the first mover (Goldman and Ross 2003, 375–79).[27] For example, in 1945, when the British Navy reengaged the Pacific theater of World War II, its alliance with the United States facilitated greater power projection through several mechanisms, including technology transfers. The British Navy then used its alliance with the United States to learn more about adopting carrier warfare.[28]

In addition to external response strategies, states may also pursue internal military changes. One option is to try to adopt part or all of the innovation, meaning the innovation diffuses from country A to country B. Partial adoption often involves adopting technological or operational but not organizational aspects of an innovation, since it is generally much less disruptive to adopt

[26] I am grateful to an anonymous reviewer for making this point clearer.

[27] The argument here makes sense even if the balancing/bandwagoning terminology is flawed, as some have observed, because states generally view themselves as balancing even if an external perception would suggest otherwise (Schroeder 1994, 119).

[28] For more on this example, see chapter 3. Essentially, alliances help boost the power of states by allowing them to combine their resources, transfer technology/knowledge, and free up resources that a state might have to spend elsewhere, among other mechanisms. For more on alliances, see Leeds, Long, and Mitchell 2000; Leeds 2003; Levy 1981; Morgan and Palmer 2003; Stam 1996.

technologies than to change the way an organization thinks about employing military force. Some states may lack the capacity to fully adopt in the short-term but will build or buy some of the technological components. This introduces a time-delay element.

Alternatively, some states might have the financial means to fully adopt and recognize that adoption makes strategic sense, but be unable to adopt for organizational reasons. This type of state might think that adoption would trigger domestic revolts or upset key military social hierarchies, outweighing the military benefits from adopting and making effective implementation unlikely. For example, in response to defeats at Ulm and Austerlitz at the hands of Napoléon's armies, the Austro-Hungarian Army reorganized itself into divisions and corps, adopting the French organizational form. Despite the best efforts of Archduke Charles, however, fears of internal instability from an armed populace caused Austro-Hungarian emperor Francis I to veto the more widespread reforms to the recruiting system required to truly adopt mass mobilization, the crux of Napoleonic warfare (Herrera and Mahnken 2003, 214–15).

Additionally, some states have the raw financial and organizational capacity to adopt an innovation, but do not do so because the financial and nonfinancial (frequently domestic political) costs of adoption are viewed as higher than those of nonadoption, especially when an external alliance strategy is possible. In the same Napoleonic warfare case, despite possessing the capacity to adopt, the British Army instead pursued an alliance strategy. Army and political leaders thought that the probability of extensive British land engagements on the continent was sufficiently low that the domestic costs, in the form of public upheaval, from shifting to mass mobilization through conscription outweighed any benefits. The existence of a viable alliance strategy made paying the costs of adoption less attractive to Great Britain (Herrera and Mahnken 2003, 216–17).

Another internal response strategy is countering an innovation, also sometimes described as offsetting (Goldman and Ross 2003, 288–89). Countering is defined for the purposes of this book as an internal military response that excludes adopting the innovation, yet can range from trying to neutralize its impact with inexpensive tactics drawn from existing forces and operational plans to the search for another new military innovation to counter the first one. An example of countering through existing military means was the camouflage practices used by the Serbian military in response to the U.S.-led air campaign over Kosovo in 1999 (Thomas 2000, 13–15). A case of a countering strategy that became an innovation in its own right was the development of the *trace italienne*, the squat and broad-angled bastion fortress. It came about as a response to increasing cannon quality that gave attacking armies the ability to batter down the tall, thin walls of most fortresses in early modern Europe (Parker 1996, 9–11).

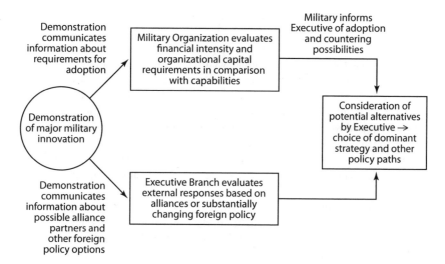

Figure 2.2. "Ideal" state response to demonstration of major military innovation.

In general, revelation of an "average" (MMI) triggers two simultaneous processes in an "ideal" state.[29] The military evaluates the requirements for adopting the innovation in relation to its current and potential capabilities, along with the possibility of a countering strategy, while the executive determines the potential utility of external policy solutions like alliances or a shift to neutrality. In states based on civilian control of the military, the military reports back to the executive, who then weighs the relative costs and benefits, probably with continuing advice from military leaders, of the different strategic possibilities. In the case of an average middle power with an advanced industrial base and medium-sized economy, the most likely choices are partial adoption, especially of some technological components, but a larger emphasis on an alliance-based response. The character of the alliance-based response depends on the relationships between the responding state, the first mover, and other potential adopters. Given the current international security environment, if the responding state is Canada and the first mover is the United States, bandwagoning is likely. If the responding state is Japan and the first mover is the People's Republic of China, balancing is probably more likely. Figure 2.2 graphically depicts an idealized version of this response process.

Given this range of options and idealized decision-making process, what factors determine how military power spreads throughout the international system? What influences the paths that particular states will adopt?

[29] Ideal types are theoretical constructs of potential state behavior useful for identifying what many states might do but not designed to exactly mimic state behavior (Weber 2002).

ADOPTION-CAPACITY THEORY

As previously discussed, it is in the response to a major military innovation that the impact on the international security environment emerges. The two key factors that determine how hard it is to adopt an innovation and whether or not a state has the capacity to adopt, which then influences the strategic pathway it is likely to choose, are the *financial intensity* and *organizational capital* required to adopt the innovation. Adoption-capacity theory posits that the financial and organizational requirements for adopting an innovation govern both the system-level distribution of responses and the way that individual actors make decisions, as well as the subsequent implications for international politics.[30]

Evaluating the range of available strategic choices for states is conceptually similar to studies of strategic choices under conditions of uncertainty in the international security environment (Fearon 1994b; Lake and Powell 1999; Schelling 1966). One difference is that adoption-capacity theory is more decision theoretic. At the system level, the theoretical predictions are based on the requirements for adopting the innovation and assumptions about the distribution of capabilities in the international system. While over time all of these things vary, in the short-term they are invariant enough to allow for stable predictions. Based on a range of possible choices, the theory also derives the most probable strategic choices for individual states given their capabilities, the requirements for adopting the innovation, and the configuration of the international system.[31]

Financial Intensity

The idea that resources matter in determining global power is not new. A core tenet of many international relations theories is the idea that material capabilities influence the overall relative balance of power in the international system, driving the alliance and other decisions of states, whether mechanistically, through the way it influences debates between interest groups within countries, or in other ways (Baldwin 1979; Waltz 1979). Financial intensity refers to

---

[30] This assumes a somewhat "normal" distribution of capabilities in the international system at the time the innovation is demonstrated. While capabilities in the international system are rarely normally distributed in a statistical sense, there is usually a sufficiently broad range of interests and capabilities such that the assumption is useful for understanding the effect of an innovation.

[31] This makes sense when thinking about the spread of military power because military organizations are not infinitely pliable, especially in the short-term. So while the theory deals with strategic choices, it does not deal with them in a game theoretical fashion that allows for mixed equilibriums. If adoption-capacity theory is incorrect, military organizations will end up being much more elastic than predicted, suggesting that future research efforts might more fruitfully focus on the issue differently.

the particular resource mobilization requirements involved in attempting to adopt a major military innovation.[32] The higher the cost per unit of the hardware associated with an innovation and the more the underlying technologies are exclusively military oriented, the higher the level of financial intensity required to adopt the innovation.

The cost per unit of technologies is important. Lower unit costs can allow for more experimentation by a military organization, making it easier to test new equipment and determine its feasibility for full deployment. For example, changes in production methods made possible by the Industrial Revolution helped lower the cost per unit of the rifle, which had existed for over two hundred years, but never systematically made its way on to the battlefield because of its previously high unit costs. Alternatively, the production of one B-2 bomber, costing between one and two billion dollars per plane on average, is so enormously expensive that experimentation is not possible on the same scale (Pike 2005a; U.S. Air Force 2006a).[33]

The underlying technology driving a military innovation can take many forms, but it is usually either an essentially commercial technology, meaning private businesses have economic, nondefense incentives to develop the technology, or an essentially military technology, meaning it was invented for military reasons and will arouse little initial interest from businesses outside of defense contractors. Underlying commercial technologies will generally require lower net capital investments for a military organization than underlying military technologies, because market competition creates incentives for private firms to pay some of the product development costs.[34] For instance, the cannons produced during the early modern European period were solely military investments, requiring a higher relative capital investment than something like the railroad in the mid- to late nineteenth century, which nondefense commercial interests had an incentive to build.[35]

---

[32] Financial intensity is somewhat related to the economics literature on "capital intensity," which refers to the level of financial investment necessary to accomplish a task in comparison with the physical or material investment in the form of labor. For some recent examples, see Arai 2003; Berman, Bound, and Machin 1998; Gartzke 2001; Sattinger 2004. In contrast, financial intensity is only partly about the overall investment level. While it is possible in the military power area that there are some innovations that require more labor and less capital to implement, and vice versa, financial intensity refers to the structure of capital investments. Nevertheless, it is certainly true that innovations requiring high levels of financial intensity to implement generally require higher capital levels, potentially necessitating a trade-off with labor given limited budgets.

[33] The type of experimentation that is more likely to result when the cost per unit is low may have interesting spillover effects on organizational capital as well. Experiments can help acclimate military cultures to new technologies and organizational practices.

[34] There may be reasons such as secrecy or control of production why the military would want to develop and produce some of its own weapons. The relative cost, however, will be higher.

[35] It is true that cannons originally developed from casting procedures used for bells, but unlike railroad tracks, once invented, there was no reason to produce cannons except for military purposes.

The financial intensity level of each innovation is derived using budget data to determine the cost per unit of the technology in comparison with the cost per unit of the prior dominant technologies, and using available data on production processes to determine whether the underlying technology is mostly commercial or mostly military. Depending on the case, the financial intensity capabilities of each potential adopter are then measured by comparing factors like the unit cost of the technology for the first mover to the defense budget of potential adopters, the gross domestic product (GDP) and defense spending of the first mover to that of potential adopters, and the relative industrial capabilities of the first mover and each potential adopter, taken from national capabilities data.[36]

Table 2.1 shows how these factors determine the relative level of financial intensity required for adoption. MMIs that require high levels of financial intensity for implementation, meaning the cost per unit is high and the underlying technological basis is military, not civilian, are harder to adopt than those requiring low levels of financial intensity.

> **Financial Intensity Hypothesis:** The greater the financial intensity required to implement the innovation, the slower the spread of the innovation at the system level and the lower the probability that a state will attempt to adopt the innovation.

Organizational Capital

Just as the financial requirements for adopting an innovation can vary, the organizational requirements for adopting an innovation can also vary. For example, in the late nineteenth and early twentieth centuries the British Navy confronted two large innovations in naval warfare. The first, battlefleet warfare, was the culmination of naval industrialization. Epitomized by the *Dreadnought*, the first all-big-gun battleship, battlefleet warfare required changes in the education, recruitment, and training of naval personnel. Instead of training soldiers to climb the rigging, for example, soldiers had to learn advanced math to plot gun trajectories. And instead of mastering sails, they had to take apart and put together engines. But the core of the innovation involved doing what the British Navy had always done: bringing its guns to bear in battle to destroy the capital ships of the enemy. This made the innovation organizationally easier to implement. The second innovation, submarine warfare, represented a more disruptive change. The potential shift in naval power away from the gun and capital ship, and toward the torpedo and submarine, contradicted centuries of thinking about the production of naval power and specifically challenged the core competency that had defined British naval supremacy: gun

---

[36] Data often drawn from the Correlates of War Project Material Capabilities data set (2006).

TABLE 2.1
Financial Intensity Drivers

| | | Cost per unit | |
		Low	High
	Civilian	Low financial intensity	Medium financial intensity
Underlying basis of the technology	Military	Medium financial intensity	High financial intensity

battles between capital ships. The innovation required not just changes in education, recruitment, and training but also wholesale shifts in force structure and plans for the use of force. This made it much more difficult to implement, and is one of the reasons that Admiral John Fisher's attempt to transition the British Navy away from the all-big-gun battleship and toward flotilla defense failed (Lambert 1995a, 1999; Sumida 1995).

Thus, different military innovations can place different demands on organizations as far as the degree and type of transformation required. Organizational capital is an intangible asset that allows organizations to change in response to perceived shifts in the underlying environment. For militaries, organizational capital represents a virtual stockpile of change assets needed to respond to changes in the character of warfare. For those who study firms, one simplistic way to measure organizational capital is to write an equation that subtracts the value of the physical assets held by a firm from the value of the company as perceived by investors, measurable in its stock price. The difference represents the intangible value of the company, or the value not captured by its physical assets.[37]

Jason Cummins (2004, 4), from the Division of Research and Statistics in the Federal Reserve System Board of Governors, provides one useful way to define organizational capital in the context of information technology:

> So what exactly is organizational capital? As a purely mechanical matter, I define organizational capital as an adjustment cost from IT investment, defined as the difference between the value of installed IT and that of uninstalled IT. Suppose a company purchases database software. By itself, database software does not generate any value. At a minimum, the software must be combined with a database and, perhaps, a sales force. Organizational capital defines how the database is used and, consequently, how software investment creates value.

[37] This can include things like brand reputation, financial accounting issues, perceived knowledge as it relates to future innovations, human capital in general, and other issues (Bond and Cummins 2000). See also Lev and Radhakrishnan 2003, 2005.

New military technology, like the new database software described by Cummins, does not produce value on its own. It is only in combination with organizational methods that technology has value for producing military force. In this way, organizational capital is the *non*technological aspect of force generation, demonstrated through doctrine, education, and training. Organizations with a high degree of organizational capital are much more able to take advantage of new innovations and transform themselves successfully for the future than organizations with a low degree of organizational capital.

The preceding discussion raises a measurement problem: if organizational capital consists of relatively intangible organizational assets, how should these assets be measured? The demands on military organizations in periods of uncertainty vary in ways potentially similar to variance in the demands on firms during periods of rapid change. Lynne Zucker and Michael Darby posit that successful firms take advantage of organizational assets and a culture of innovation to stay ahead during periods of uncertainty. Firms, even leading ones, are not necessarily immutable. For example, the fact that some leading pharmaceutical firms radically shifted their identities in response to the biotechnology revolution while others failed demonstrates that different firms respond in different ways to external demands (Zucker and Darby 1996a, 960, 971; Zucker and Darby 1996b).

Organizational capital is a critical element of firms' success, helping businesses survive during periods of change by giving them the institutional capacity to shift practices without being blocked by bureaucratic obstacles. Because it is difficult to copy business processes, as opposed to specific technologies, organizational assets are difficult to duplicate. Darby and Zucker find a ten- to fifteen-year gap between the beginning of the biotechnology revolution and widespread knowledge of the practices necessary to take advantage of the revolution for business purposes. In between, tacit knowledge of how to conduct specific research was held by key scientists working for a few firms and others in their immediate vicinity, making the knowledge "naturally excludable" from outside firms (Darby and Zucker 2003, 2). These human resources then became the basis by which an existing incumbent firm stayed ahead when further change appeared likely (Zucker and Darby 1996a, 12709; Zucker and Darby 1996b, 965–67).

Even detailed case studies on business processes have frequently been insufficient to enable copying by other firms. Erik Brynjolfsson, Lorin Hitt, and Shinkyu Yang (2002, 144) explain the difficulties involved:

> Intel, for example, has adopted a "copy exactly" philosophy for any chip fabrication plant built after the first plant in each generation. Wholesale replication of even seemingly insignificant details has proved more reliable than trying to understand which characteristics really matter. Going from the plant level to the firm level only complicates the imitator's task.

A useful set of alternative operationalizable markers from this business innovation literature can help us measure both the level of organizational capital possessed by a military organization and the baseline level of organizational capital required to implement a given MMI. Three measures of the level of organizational capital possessed by a military organization are the critical task focus of an organization, experimentation resources, and organizational age.

CRITICAL TASK FOCUS

Typically measured by statements of intent and planning documents, the critical task of an organization is that which the organization views as its main "goal"—what it seeks to achieve. While not always referred to as a critical task in literature on organizational effectiveness and the integration of innovations, the notion of a critical task captures the central operating principle of organizations. Having a critical task helps organizations frame and justify their actions, providing a central theme for motivating workers. As business innovation scholar Clayton Christensen (1997, 168) writes, "Over time, however, the locus of the organization's capabilities shifts toward its processes and values." Critical task focus is the extent to which the means by which an organization achieves its goals are conflated with the goals themselves (Wilson 1989, 25–26). Many successful firms come to define their business processes and values based on the strategies that initially led to success. The embedding of a particular vision of a critical task within the organizational architecture then shifts from where a business allocates resources to the metrics used to make decisions to the processes used to design metrics. One example is the failure of the Digital Equipment Corporation (DEC) in the shift from minicomputers to the personal computer. The organizational process at DEC prioritized reinforcing the high-margin minicomputer business, even though it also had high overheads. Since the emerging personal computer market looked miniscule in comparison, with low overheads and profit margins, DEC's organizational decision making blinded it from seeing the shift in the market. As Christensen (1997, 169–71) describes, eventually DEC failed, not because it lacked the resources to compete in the personal computer arena, but because the business processes designed to maximize profits in the minicomputer arena embedded an inappropriate critical task at the core of the business.

In general, businesses with narrow critical tasks associate particular ways of doing business with the goal of the business. Businesses with broader critical tasks are those that delink the goal of the organization from the means used to achieve that goal. Kim Clark shows how firms often use their experiences to build specific types of knowledge and skill. They identify key problems and solve them through the creation of efficient business practices. Still, when exogenous technological shocks occur or customer preferences change overnight, successful businesses can find themselves hamstrung by the organizational climate they previously created (Clark 1985). Competence-destroying innovations, like the development of private airlines or mass production processes for cement, can disadvantage firms that previously held large segments of market share (Tushman and Anderson 1986).

The critical task focus for a firm thus becomes myopic at times not because a firm is poorly managed but precisely because smart managers optimize their entire organization, from production processes to decision-making hierarchies, to grow their market share in a particular business area (Christensen 1997, 98). Firms frequently create strategic group identities to reduce uncertainty through the use of predictable metrics and planning processes. Especially when these identities form around production processes or specific ways of achieving market share, they make it more difficult for firms to respond to radical innovations in core practices (Henderson and Clark 1990).

In contrast, the broader the critical task focus—meaning the less specifically a firm links its organizational identity to particular production methods—the easier it is to incorporate new ways of achieving organizational goals—that is, adopting an innovation—without encountering too many bureaucratic stumbling blocks to implement the change. The more specifically a military organization defines its critical task, the harder it should be for the military to adopt an innovation. Entrenched interests within the organization will be more likely to rebel on the grounds that a proposed innovation is outside the scope of acceptable activities. An organization with a broader critical task focus will find it easier to justify, within the organization, how the innovation fits within the regular "actions" of the organization.

One often-cited example of differences in critical task focus comes from the U.S. Marines and U.S. Army. The critical task of the Marines is described broadly as being a "warrior," and the Marines envision all Marines as soldiers with rifles first rather than identifying each Marine by their specific task. This has made the Marines more receptive to a wider variety of core operational concepts including amphibious warfare, trench warfare, counterinsurgency, and urban warfare, since they all involve the activities of a warrior. The army, in contrast, as James Q. Wilson (1989, 43–44) has pointed out, has traditionally defined its critical task as massing firepower to win conventional wars, making it more difficult to justify the incorporation of new war-fighting methods like counterinsurgency (Krepinevich 1986, 5). The army preference for operations emphasizing overwhelming firepower also likely influenced the army's decision to use casualties inflicted on the enemy as its dominant indicator, or key metric of success, during the Vietnam War (Gartner 1997).[38] The army's operational focus on overwhelming firepower drove its strategy, rather than the other way around. This is related to work by John Nagl (2005, 216–17), who

---

[38] Beginning in 1965, General William Westmoreland based U.S. strategy around search-and-destroy missions designed to use overwhelming firepower to generate casualties (Gartner 1997, 130–31). This explanation is very much in line with adoption-capacity theory.

uses the U.S. Army in the Vietnam War to argue that the way organizations view themselves and their critical mission can hinder learning even in the face of strong evidence suggesting current methods are failing.

The prescriptions for firm success outlined by Christensen use the organizational commitment to experimentation as an accurate way to measure organizational capital. Christensen contends that the firms best prepared for disruptive innovations are those with ongoing experimentation efforts and/or spin-off companies that can avoid substantial losses if the spin-off fails. Spin-offs with loose ties to the established company, because they measure success in relation to their new competitors, rather than the rest of the company, are best positioned to highly value the small profit margins likely to accrue at the beginning of a transformative period (Christensen 1997, 176–77).[39]

High overall investments in experimentation may make militaries more receptive to innovations; experiments demonstrate an institutional willingness to think outside the box, meaning the organization builds in the capacity to integrate innovations. The George W. Bush–era Office of Force Transformation in the U.S. Defense Department was one example of a subunit focused on experimentation. Within the Marine Corps, the Marine Corps Warfighting Lab is another locus of experimentation. In fact, while a key finding in the business innovation literature is the relationship between corporate "skunkworks" and success in the face of innovation, the term itself is derived from the work of a defense contractor. When Lockheed designed the P-80 Shooting Star jet plane during World War II under high secrecy, a circus tent with a top-secret incubator was positioned next to a plastics factory. Rogers (2003, 149) writes, "The strong smells that wafted into the tent made the Lockheed R&D workers think of the foul-smelling 'Skunk Works' factory in Al Capp's Li'l Abner comic strip. The name stuck and came to be generalized to similar high-priority units that have been created by various companies since."[40]

Organizations generally acquire some degree of rigidity over time, independent of domestic politics and overall political centralization, throwing up barriers to transformation that undermine organizational capital levels. While business innovation research demonstrates that firm identities are not entirely immutable, the conventional wisdom—that it is difficult for large organizations

---

[39] Because the market size is smaller for emerging component markets, the profit margins are often smaller.

[40] Experimentation is a better proxy than research and development. Organizations may pump tons of resources into research and development, but if they do not invest in the right areas, they may end up promoting incremental improvements to the *last* great thing, rather than the next great thing (Christensen 1997).

to transform to face new business challenges, especially when the challenge is disruptive to their fundamental business model—is the conventional wisdom for a reason. Olson's research shows that the age of an organization often predicts its productivity level.[41] Based on a number of case studies and a wide variety of data from studies of businesses to studies of cross-national post–World War II economic growth, Olson's work (1982, 76–77) demonstrates that older organizations many times become more stagnant, making them ill equipped to deal with transformational change.[42]

Olson (1982, 31) argues that collective action incentives decrease as group size increases, because the benefits from collective action are more diffused, making it harder for larger groups to act in pursuit of common interests. Attempts to solve these problems by creating subgroups to promote change often backfire, fueling the development of bureaucratic sublayers of inefficiency by reformers and corresponding interest groups that mobilize to block reform, requiring special interest groups to help people navigate the bureaucracy. The result is an increasing number of veto points in the overall organization with the capacity to block change. Interest groups that have cartelized for the purposes of distributing benefits will have incentives to block resource redistribution, especially to new groups forming in response to changing conditions. This can delay the adoption of new technologies and the more efficient distribution of resources (Olson 1982, 64–65). The theoretical extension to the military realm is obvious. Well-defined military service cultures often militate against change, because service groups will defend existing distributional bargains about the spending of defense dollars against attempts to reallocate. The basic prediction is that as organizations age, the level of organizational capital declines.[43]

These three measures of organizational capital are potentially operationalized by using internal documentation (when available) to code for the critical task focus of militaries, budget data to code for the commitment to experimentation, and the time gap since a regime change or a major war in which a

---

[41] For some critiques of Olson, see Coates and Heckelman 2003; McLean 2000; Unger and Van Waarden 1999.

[42] For example, Olson argues that the destruction of Germany and Japan during World War II allowed the two countries to start anew economically, resetting their organizational ages. This made it easier for them to take advantage of innovations in the post–World War II period (Olson 1982, 76–77).

[43] There are additional indicators likely unrelated to the transformational capacity of organizations but that may matter, including the centralization of authority, receptivity to information, civil-military relations, and organizational size (as opposed to organizational age, which is more about whether the bureaucracy is bloated than its absolute size, though they are sometimes related). The importance of centralization, civil-military relations, and information are probably subsidiary questions to the issues of critical tasks and organizational age, since the predictions are relatively similar (i.e., the expansion of veto points as organizations age induce the same types of dysfunctions as those predicted by a civil-military explanation). The question of organizational size likely depends on the character of the military innovation and varies based on other factors.

state lost, whichever occurred more recently, to code for organizational age.[44] When evaluating the organizational capital requirements for a given innovation, I determine the level of organizational capital possessed by the first mover(s) for each innovation, which is a good proxy for the requirement since there is sometimes uncertainty about the exact measurement of the three indicators. Combined with general measurements, a given innovation is assessed as requiring a certain amount of organizational capital to adopt.[45] Then, the organizational capital required to adopt an innovation is compared with the organizational capital possessed by individual states to determine the propensity of each state to adopt the innovation.

Organizational Capital Hypothesis: The greater the organizational capital required to implement the innovation, the slower the spread of the innovation at the system level, and the lower the probability that a state will attempt to adopt the innovation.

It is important to note that this understanding of organizational capital is not without its limitations. Studying organizational change is difficult, and the wide variety of factors that determine the way organizations make decisions makes it impossible to encompass all of them in any particular theory. The three factors described above—critical task focus, experimentation, and organizational age—are designed to proxy for the type of characteristics likely to exist in organizations more or less able to adopt new military innovations.

Predicted System-Level Diffusion Pattern

While the particular international context in which an innovation arises matters a great deal, the best general predictors of the way it is likely to diffuse throughout the international system are the relative levels of financial intensity and organizational capital required to adopt the innovation. Combining the predictions about financial intensity and organizational capital highlights the way that different requirements for adopting an innovation should influence the way it spreads throughout the international system.

[44] When possible in the empirical chapters, the regime change data comes from Polity while the war loss data comes from the COW Project (Marshall and Jaggers 2002; Sarkees 2000). Major wars are measured as wars where the demand involved regime change (Horowitz, Simpson, and Stam 2009). An alternative measure is wars that are followed by regime change, though this conflates the two measures.

[45] Using the first mover as a model for the required level of organizational capital is imperfect, but it does contextualize the adoption requirements. It is possible that over time, the level of organizational capital required to adopt some innovations will decline. This is an important limitation to the first-mover test. Yet it is only one of several ways of measuring organizational capital. If the first-mover test leads to an exaggeration of the adoption requirements, it will show up in the empirical chapters, since the adoption-capacity theory will predict a much smaller and more limited spread of the innovation than what actually occurs.

TABLE 2.2
System-Level Diffusion Predictions

		Level of financial intensity required to implement major military innovation	
		Low	High
Level of organizational capital required to implement major military innovation	*Low*	Rate/extent fast	Rate/extent medium
	High	Rate/extent medium	Rate/extent slow

State-Level Predictions

Moving from system-level predictions to individual states requires acknowledging additional complexity. One key factor is whether knowledge concerning how to adopt has diffused to a particular state. While the adoption-capacity requirements can predict how an innovation will spread in general, forecasting the decisions of specific states requires understanding their particular motivations as well. One could argue that the system-level predictions, when applied to individual states, are tautological because they predict that those states with the capacity to adopt will do so. Yet threats and internal politics may lead states toward choices that maximize their utility for preference functions other than making the most of short-term military power.

STRATEGIC CHOICES

In general, states responding to a major military innovation are likely to choose a dominant response strategy that consumes the bulk of their time and effort from the range of possible choices outlined above—though they may simultaneously pursue a few different paths.[46] The dominant strategy is an attempt by the state to maximize its utility in response to a given innovation, though states can and do make bad choices. The payoff for adopting a particular strategy is a function of state certainty about the substance of the innovation, depending on how effectively information about the innovation has spread, the interest a state has in adopting given its geopolitical position and domestic political configuration, and its adoption capacity, its ability to adopt the innovation based on the combination of its financial and organizational resources. While other factors also influence responses to innovations, these factors should generally be the most critical.

As Posen's work (1984, 233, 239–40) on the interwar period in Europe suggests, large-scale geopolitical threats could increase the urgency with which

---

[46] In an idealized state, while initially states will have debates over the appropriate response to an innovation, with advocates existing for a range of possible responses, eventually one emerges for a variety of reasons as the dominant response even if elements of others continue to exist. Alternatively, see Resende-Santos 2007, 67–69.

states respond to new military innovations. Threats will not necessarily lead to adoption, however—just the optimization of response strategies. Depending on its preinnovation configuration of alliances and adoption capacity, a state's optimal response strategy could be a dominant alliance strategy, an adoption strategy, a countering strategy, or a shift toward neutrality.

Domestic political costs and benefits may also influence state decisions. For example, the failure of the Archduke Charles to fully implement the Napoleonic reforms in the Austro-Hungarian Army, discussed above, was rooted in domestic political turmoil that prevented effective mobilization, and meant that if mobilization did occur it would threaten regime stability. So despite a strategic need and the financial capacity to adopt, a lack of organizational capital and domestic political constraints prevented successful adoption.

In situations where geopolitics and domestic politics influence response strategies away from adoption, it becomes more likely that states will pursue countering strategies as one alternative. Countering attempts, which try to offset the value of an innovation through low-cost alternatives or substitutions for a lack of particular capabilities, are generally much easier for an organization than adoption. The potentially inexpensive nature of countering may make it an especially attractive solution for states that may believe they can emulate and adopt the innovation over time but need to stall in the interim, or states lacking the capabilities to adopt in any reasonable time frame but that do not want to abandon their current foreign policy through new alliances or a shift toward neutrality.

SUCCESS AND FAILURE

Predicting the likely dominant strategy that states will choose in response to changes in the production of military power is not enough to answer the puzzle posed at the outset of this book. Determining how military innovations diffuse requires figuring out not just what a country will try to do but also whether it will actually work. For the external strategies possible in response to an innovation, predicting success and failure is not difficult; it depends on the willingness of states to sacrifice foreign policy goals, and who their friends and adversaries are in the international system. The failure of states' countering or adoption attempts are much more interesting. The requirements for implementing each innovation determine which states are least likely to adopt and/or most likely to fail, and these vary widely across innovations. This builds on Goldman's notion of capacity and distinguishes adoption-capacity theory from realist perspectives, given that Resende-Santos (2007, 9–11) explicitly notes that neorealist theories like his cannot explain the success or failure of strategies, only whether states will emulate.

For those states that attempt to adopt an innovation, the key determinant of success is whether the financial and organizational requirements for implementing the innovation "match" the capabilities of the state pursuing an adoption

strategy.[47] These claims about success and failure could function in two different ways: one focusing on selection, and the other focusing on organizational
myopia. A selection argument might presume that a state only attempts adoption if it is the best option, meaning that the state has the capacity to successfully implement the innovation.[48] On the other hand, as Robert Jervis (1976,
189–91) maintains, slowly changing perceptions can cause organizational myopia as actors fail to correctly assess their own capabilities and the situations
they face. The empirical tests in subsequent chapters will help arbitrate between these two different ways of thinking about the relationship between
strategic choice and implementation success.

Implications for International Politics

Given that innovations represent potentially important shifts in military power
and that different innovations are likely to have different adoption requirements, influencing the way that they spread in the international system, what
does that mean for the international security environment? One way of thinking about the impact of innovations is to look at the content of particular innovations and the way they influence international interactions. For example,
since nuclear weapons radically increased the destructive power of a single
bomb, they changed the character of warfare. In another possible case, the
"modern system" took advantage of changes in firepower, transportation, and
communications, leading to large shifts in force structure and battle planning
that escalated the destructiveness of wars (Biddle 2004). But this type of analysis alone would view war too much in a vacuum, rather than evaluating the
relative implications of innovations for the international security environment.
While the content of specific innovations clearly matters for changing the way
a particular war is fought, it is the spread, or lack thereof, of those innovations
that determines the overall winners and losers in international politics.

   If military power is based in the ability to adopt the key military methods of
a period, then the creation of military innovations and their disparate impact
on different states throughout the system is one way that large-scale shifts in
the relative military balance of power occur.[49] Power transitions, or periods

---

[47] Adoption capacity is clearly endogenous in many cases to strategic choices; states with higher
financial and organizational capabilities will be more likely to attempt adoption in most cases,
though the British response to Napoleonic warfare, described above, shows that this is not necessarily the case.

[48] It is also possible rationally that an attempt to adopt an innovation might be the preferred
strategy even if leaders believe the attempt will fail. The attempt could signal a commitment internationally that influences allies and potential adversaries, and might also just be the least bad option in a comparative sense, given the relative costs of the various external policy alternatives.

[49] This builds on Gilpin 1981, 59–62.

when some states rapidly rise while others decline, are critical times in global politics because they may be the dry timber through which major fires, large-scale wars, can start.[50] Analyzing how the rise and decline of states influences the international security environment has been an important subject in international relations for decades (Copeland 2000; Gilpin 1981; Kugler and Lemke 1996; Organski and Kugler 1980; Waltz 1979).[51]

Understanding the diffusion pattern of the dominant innovation of a time period and the match of adoption requirements to the capabilities of states can help us more accurately explain power transitions. In contrast to his general contention about the inevitable decline of existing powers, Robert Gilpin (1981, 161–62) writes, for example, that existing powers can reinvent themselves and stay on top, pointing to Chinese and British developments over time. He does not explain, though, why it is that they can reinvent themselves when facing some challenges, but face "institutional rigidities" (189) at other points that hasten their decline. So what explains their eventual decline?

Most narrowly, adoption-capacity theory can explain the *mechanism* by which these transitions sometimes occur. Military innovation diffusion is part of the causal process governing power transitions.[52] The predictions made by adoption-capacity theory about the way adoption requirements and state responses will influence the spread of a military innovation also lead to predictions concerning the importance of innovation diffusion for the international security environment. Further, differentiating between financial and organizational requirements for adopting innovations and the implications for international politics can also explain the process by which new states take the lead for Gilpin, or how the costs of war shift for Powell.

More broadly, at the system level, adoption-capacity theory explains both why some shifts in relative power occur and how. Some new major military innovations constitute disruptions in international politics that can generate large power disparities. First movers that innovate and generate new ways of producing military power can gain significant advantages; the exploitation of these advantages then can usher in power transitions, exposing status quo powers that can become overmatched paper tigers. Rising powers that become first movers are especially likely to experience large gains in relative power. First-moving existing powers or existing powers that can rapidly emulate or

[50] The phrase power transition is used to describe increases and decreases in relative power, not the power transition "school" (Organski and Kugler 1980). Subsequent discussions of shifts in power should hold whether or not traditional power transition dynamics or Powell's argument (1999, 85, 199) about the divergence between the distribution of power and the distribution of benefits holds.

[51] In this section, the terms existing and rising power refer strictly to relative capabilities in the period prior to the introduction of the innovation, not intentions.

[52] For an example of work that does discuss the affect of innovation diffusion on power, see Goldman and Andres 1999.

otherwise adapt to an MMI can make relative power gains, or help stave off disadvantageous power shifts. Innovations more generally trigger strategic responses on the part of the rest of the international system, with success and failure in the world of the new innovation determined by the match between the financial and organizational requirements for adoption and the financial and organizational capabilities of interested states, modified by their alliance postures and other options.

New major military innovations sometimes present major powers with challenges they simply cannot meet, if they lack the financial or organizational capacity to adopt. These shifts in relative power and periods of instability with declining powers do not occur because a state makes a mistake. Rather, sometimes the financial and organizational requirements for the next major military innovation are just beyond the capability of a major power to implement. Adoption-capacity theory is a protorationalist explanation for change in response to innovation; rather than looking to pathologies or other reasons why states fall by the wayside in the international power game, it is simpler to look at what it would have taken for a state to "keep up," and whether the state was simply unable to do so. Especially when preexisting powers desperately want to stay on top, they may end up with an attribution bias that causes them to exaggerate their capabilities. It is in these situations that the most dangerous power transitions occur. These faltering states are the most likely to engage in foolhardy military strategies that make escalation more probable.

Alternatively, the recognition that their relative military power will only decline for the foreseeable future could create incentives for risky military gambles as leaders enter a "use it or lose it" mind-set. Fear of structural military decline could also cause a general heightening of tension that leads to war as a declining power covers for insecurity about their capabilities with greater levels of belligerence at the same time that potential adversaries are gaining military confidence due to successful adoption of the new innovation. Strong incentives might exist for national leaders to try to gamble for resurrection with regard to the reputation of the country as a global power, especially given the potential personal cost to the leader if the leader is blamed for overseeing national decline (Goemans 2000).[53] While this possibility exists, it is also extremely difficult to test since it requires not only knowing what a country did and why but getting detailed information on the risk perceptions of leaders as well.

Rather than decline being inevitable, as it is for those that presume the diffusion of knowledge will inevitably harm system leaders, shifts in power and their timing are better understood as a function of the adoption requirements of the key military innovations of a period.[54] Not all innovations will benefit rising

---

[53] On the other hand, the risk of removal from office if the state gets involved in a war and loses could induce caution.

[54] Ronald Rogowski (1983, 735) makes this point about Gilpin's inevitability argument as well.

powers and hurt established ones. An international configuration that might seem ripe for change could hang on for decades or longer because the dominant form of military force favors the existing power. It is only the emergence of a new innovation requiring a high level of organizational capital to adopt that drives a nail through the coffin of the existing power. Or as Ronald Rogowski (1983, 735) hints, it is necessary to combine the organizational insights of Olson with Gilpin's arguments about shifts in relative power, which adoption-capacity theory does. This allows us to predict not only that a shift in the distribution of power is likely but also the *timing and character* of that transition.

So how do different adoption requirements for innovations, when summed over the international system, actually translate into changes in how an innovation will influence the international security environment? There are two key ways that the diffusion of military power matters for the overall international security environment: through the implications of the system-level distribution of responses, and from first-mover advantages and power asymmetries for particular states. The former is more about the relative balance of power between states, while the latter is more about the implications for the relative timing, intensity, and geographic scope of warfare.

DIFFUSION AND SYSTEM-LEVEL CONSEQUENCES

Returning again to the business innovation literature, Christensen highlights the way that differences in the adoption requirements for innovations in the business world influence the probability of firm success and failure. He differentiates between sustaining innovations, or those that improve current business practices, and disruptive ones, or those that make following prior business practices likely to lead to firm failure instead of success (Christensen 1997, xiv–xv).[55] Christensen studies the disk drive industry and its transition through several different stages of innovations. He finds that in every case of a sustaining innovation, like thin-film disks and heads, established firms like IBM led the way. Yet disruptive changes almost always caused major transitions in the industry. By the time the new technology became clearly more efficient than the old one and generated new markets of sufficient size for big firms to become interested, it was too late for entrance by most of the established firms (13–15). Essentially, leading firms failed when confronted with innovations that required a new way of doing business, but succeeded when confronted with innovations that enhanced the previous way they did business.

That insight forms the basis of the relevant predictions about MMIs.[56] Focusing on variations in the financial and organizational requirements for adopting an innovation yields four ideal types of cases: requirements are either

---

[55] These ideas build on some of the concepts contained in Joseph Schumpeter (1942).

[56] Others have also used Christensen to make arguments about military innovations, including Dombrowski and Gholz (2006, 18, 26–29) and Terry Pierce (2004).

high financial–high organizational, low financial–low organizational, high financial–low organizational, or low financial–high organizational.

Innovations requiring low levels of financial intensity and organizational capital to adopt are not terribly interesting from a relative power perspective. Nearly all affected states should be able to quickly and efficiently adopt the innovation, since it presents neither significant financial barriers to entry nor extremely complicated organizational procedures. One example of this might be the rapid adoption of chemical warfare by the entente powers following Germany's use of poison gas in the Second Battle of Ypres in World War I. Other possibilities are more interesting. MMIs requiring a large financial investment but not much organizational transformation are likely to benefit pre-innovation great powers and/or the system leader, since they are the wealthy states, and best positioned to simply build or buy the relevant technology and plug it into their military. Preexisting major powers will merely incorporate the new technologies and some doctrinal changes into their existing organizational structures, meaning they can rapidly build up in the new area of capabilities. Nonmajor powers will face huge financial barriers to entry even if adoption is possible from an organizational perspective. An example of this, discussed in greater detail below and in chapter 4, is the spread of nuclear weapons, especially in the first decades after World War II.

In contrast, innovations requiring high levels of organizational capital to adopt but low levels of financial intensity can generate disruptions in the balance of power due to the metamorphic organizational shifts required to take advantage of the new ways of war. Rebecca Henderson (1993, 250, 254–55) divides innovations into categories based on the relative degree of economic/technological and organizational change necessary for implementation, studying innovations in the photolithographic alignment equipment industry, a subset of the semiconductor industry.[57] She finds that the types of innovations that caused the most industry turnover were technologically incremental but organizationally radical, such as contact printers and scanning aligners. The failure of firms like Parker-Elmer when faced with organizationally radical changes like stepper aligners highlights the primacy of organizational change as a causal variable (261–62). Organizational inertia made the efforts of leading firms less effective than those of market entrants. While leading firms were not absent when radical innovation occurred, their ability to take advantage of those types of innovations was lower than those of entry firms. When innovations were organizationally incremental, however, leading firms capitalized because their existing business structure was reinforced by the new innovations, giving them a larger advantage over smaller firms (267–68).

---

[57] Photolithographic aligners are critical elements in the production of solid-state semiconductors, representing up to 30 percent of the production costs in a semiconductor facility (Henderson 1993, 256).

When new military innovations require high levels of organizational capital to implement, Henderson's research and Christensen's work on firm failures suggest that just as those firms that are most successful in existing markets often flounder in the face of disruptive innovations, having the ability to heavily and successfully invest in the military capabilities that enhance combat effectiveness during one period may handicap the capacity of a military organization to transform to deal with the next era. These firms struggle for similar reasons to those described by Jervis when assessing how decision makers integrate new information. Jervis explains how preexisting beliefs shape the way that decision makers integrate new information. Their worldviews, which have served them well in the past, can blind them to changing circumstances, since they see things through the lens of their past experiences and beliefs. Therefore, they apply the same cognitive map to a new circumstance even though it is now inappropriate (Jervis 1976, 143).

In the military context, since these types of innovations will require doing business differently, rather than doing it better, research and development segments of leading military organizations are more likely to dismiss early signs of the innovation as irrelevant to core competencies, placing them at a tacit knowledge disadvantage if/when they attempt to catch up later.

The low required financial intensity will decrease the barriers to entry in the international system, potentially allowing new states to compete by at least partially wiping the military slate clean. This will threaten the position of existing powers as some likely try to resist the change, while others recognize that an adoption-capacity gap means the state must choose an alternative strategy and accept a period of relative decline. Many argue that combined arms warfare, also known as blitzkrieg, falls into this category. As Ernest May (2000, 476–80) and others have pointed out, the other armies of Europe possessed much of the same, if not more, raw military equipment: radios, airplanes, tanks, and motorized vehicles. It was the way Germany put together those technologies into maneuverable, deep-strike packages that was inventive. The German departure from the French strategy of static defenses and trench warfare required a significant shift in training and operations, making maneuver warfare difficult to adopt for many militaries (Boot 2006, 233–36). France, in contrast, represented a military with a low level of organizational capital due to its high organizational age; this is one reason that its victory in World War I locked in flawed lessons about future wars that prevented it from grasping the potential of maneuver warfare (May 2000, 449–50).

Finally, shifts requiring high levels of organizational capital and financial intensity, precisely because their spread rate and extent is likely to be constrained, should theoretically have the largest effect on the balance of power because countries that do take advantage of the MMI will gain long-term asymmetrical power advantages. Major powers that lack the capacity to adopt the new MMI are also the states more prone to making strategic errors when facing a challenge.[58] The carrier warfare case, described in more detail in chapter 3, exemplifies this rare but powerful high-high case.

Table 2.3 outlines the idealized relationship between the diffusion of a given MMI in a normally distributed international system (an important qualifying assumption) and its impact on global power.

Regardless of one's theory of change in the balance of power, the above examples demonstrate that shifts in the ability to produce power matter. Military innovation diffusion represents one causal process governing the timing and character of power shifts that have been previously explained on the basis of other factors. The degree to which leading countries take advantage of new forms of military power helps predict long-term trends relating to the international balance of power.

First-Mover Advantages

Economists have argued about the relative advantages a firm achieves when it is the first to introduce a new technology or marketing strategy into a given market, making it the so-called first mover (Bain 1956; Schumpeter 1942). While the notion of first mover generally refers to the ability to bring a new product to market in a way that captures large amounts of market share, it is useful to break down first-mover advantage into its component parts. There is both a technological and an organizational component to first-mover advantages. The technological advantage refers to hardware; firms break ground directly in the area of a given product, like semiconductor improvements, or introducing entirely new products, such as the personal computer. The organizational advantage is when firms take a given technology, and figure out how to produce and market it in ways that give the firm an advantage. The production line, which Henry Ford developed for the Model T, exemplifies an organizational first-mover advantage.

There are some clear advantages to being a first mover in terms of long-term rents and business viability (Milner and Yoffie 1989, 244). One analogy between the military and business world comes from patent law, which influences the ability of firms to maintain a monopoly on production knowledge and extend their first-mover advantage (Mykytyn, Bordoloi, McKinney, and Bandyopadhyay 2002, 1). Industries like pharmaceuticals that depend on secrecy regarding production processes have shown that patents are critical to the long-term viability of first-mover advantages (Lieberman and Montgomery 1988, 43–45). Military organizational processes are somewhat (though not entirely) comparable to the special knowledge a company has when it invents a product.

---

[58] These states may or may not be in decline prior to the onset of the innovation. Adoption-capacity theory is agnostic on that point.

TABLE 2.3
Relationship between the Spread of Military Power and the Balance of Power

		Level of financial intensity required to implement major military innovation	
		Low	*High*
Level of organizational capital required to implement major military innovation	*Low*	• Diffusion fast • Impact short-term and not likely to be structurally significant • Relatively shorter first-mover advantage • Key example: chemical warfare	• Diffusion medium • Risks reinforcing existing global power balance • Relatively shorter first-mover advantage (except in extreme cases) • Relatively benefits existing powers • Key example: nuclear weapons
	High	• Diffusion medium • Risks reordering global power balance in ways likely to threaten existing powers • Relatively longer first-mover advantage • Key examples: suicide terrorism, Napoleonic warfare, combined arms warfare (blitzkrieg)	• Diffusion slow • Likely to have substantial and long-lasting disruptive impact on balance of power • Relatively longer first-mover advantage • Key example: carrier warfare

Like businesses, MMI first movers have a natural incentive to maintain secrecy about new technological capabilities and their organizational processes, since either or both could be based on knowledge that the first mover wishes to prevent from leaking. But once a demonstration occurs or enough information spreads, it is difficult to maintain secrecy for all but the most complex innovations. Like firms choosing research and development strategies, military organizations have to determine whether to allocate limited resources toward reinforcing competence in current military areas, or potentially becoming a first mover in new military technology or innovative organizational processes (Lieberman and Montgomery 1988, 54).

The length of the first-mover advantage for a given innovation affects its consequences for international politics for two reasons. The relative power of a first-mover state should increase faster vis-à-vis peer states during its first-mover period than in any other one. Additionally, the substantive effect of a specific military innovation should also be especially pronounced for first movers. For example, some innovations, like Prussian open-order tactics in the late nineteenth century, involve taking advantage of breakthroughs in transportation, such as the railroad. In that case, the content of the innovation, combined with the first-mover advantage, produced an enormous military edge for Prussia against France in the Franco-Prussian War. But the innovation quickly—albeit partially—spread, to some extent negating Prussia's first-mover advantage over the medium-term (Herrera and Mahnken 2003, 226).[59]

The adoption requirements for an innovation influence the length of the asymmetrical advantages that adopters receive, especially first movers, changing the way an innovation matters for the security environment. If an innovation is relatively easy to adopt, an actor will be unable to hold on to its first-mover advantage long enough to establish a dominant monopoly; adoption will be relatively easy for both early adopters and those that follow. The length of the first-mover advantage will decline, and the *relative* impact on the international balance of power will also be smaller. Other states in the system will mimic the first mover quickly enough to negate many of the advantages to being first.

In contrast, innovations that are more difficult to adopt, especially in terms of organizational requirements, will provide greater first-mover advantages to early adopters, helping them achieve larger international political advantages. Innovations based on new knowledge are more difficult to adopt, either because expertise at prior key competencies makes acquiring the new knowledge harder or because they are breakthroughs in new areas where others lack knowledge. These innovations are also entry enhancing, meaning the door is open for massive market upheavals (Darby and Zucker 2003, 2, 8; Tushman and Anderson 1986).

> **First-Mover Gains Hypothesis:** First movers should experience relative gains in power proportional to the length of their monopoly over the MMI and its relevance in international politics. The length of the first-mover advantage will be inversely proportional to the diffusion rate of the innovation.

First-mover advantages might not apply in all cases, however. Late movers have an edge in some business environments because they can take advantage of technological improvements to innovations that reduce the uncertainty and risk associated with a given product (Lieberman and Montgomery 1988, 47–48; Tellis and Golder 2002).[60]

---

[59] It is also important to distinguish between those states that follow the first mover, like Germany with all-big-gun battleships, and true late movers, like North Korea with nuclear weapons. "Late-mover effects" refers to the latter, not the former.

[60] This concept is related, though not quite the same, as the argument that Gerschenkron (1962) makes about the advantages of "backwardness." Gerschenkron's assertion is more about development than about the response to innovation winners. There is a potential disadvantage to being a first mover in a military area that relates to organizational inertia as well. For example, automobile

Late-Mover Gains Hypothesis: Late adopters will face lower barriers to adoption due to more available information about the innovation, giving them a relative power edge over first and early movers once adoption occurs.

War

The diffusion pattern for major military innovations can also influence a broad range of factors related to how states select into wars and fight. Rapid changes in military effectiveness may also affect the overall utility of military power as a method of accomplishing political objectives.

GEOGRAPHY

The effect of geography on international interactions, especially military interactions, is well-known (Diehl 1985; Vasquez 1995). While most disputes are between contiguous states because they are fighting to control territory, the technology and thinking about the use of military force undergirding most military operations should not be ignored. If it takes years and/or decades to reach the land of an adversary, it will be much harder to fight a war than if an adversary is right next door.

MMI diffusion, presuming the MMI is successful (which is not necessarily the case), should substantially alter the geography of military incidents by allowing states that experience MMIs to militarily interact with a broader geographic range of states. The effect should be more pronounced when both states in a dyad have experienced an MMI, but should also exist in an asymmetrical relationship.

Even if a particular innovation is not directly related to the technology of travel, major military innovations may give a state the ability to conduct more deadly military interactions in a wider geographic range. For example, the rifle and machine gun allowed for the projection of lethal force with fewer troops than were previously necessary, increasing the geographic scope of expeditionary forces.[61]

To the extent that MMIs shift outward the geographic range of the possible in a military sense, they should introduce more systemic instability, meaning

---

first-mover Henry Ford continued to use the same business model to produce his cars even after developments in the industry should have caused changes (Abernathy and Wayne 1974, 109). So the organizational model that led to his success also led to a form of bureaucratic lock in due to its high organizational age and narrow critical task focus.

[61] For instance, take two states that have normal economic and diplomatic relations but cannot interact rapidly due to geographic constraints. If an MMI makes it possible to effectively project military force faster at a much greater distance than was previously possible, one or both states will have the ability to more credibly make military threats. This could fundamentally shift state-to-state relationships.

periods with MMIs are more likely to have more militarized interstate disputes (MIDs) and wars on a systemwide level, not just for those states experiencing an MMI or competing with a state experiencing an MMI.

SPEED

In addition to widening the geographic range of states with which a given state can interact, some MMIs may also enhance the speed of those interactions. If states can more quickly mobilize military forces, they can issue credible threats more often and in a wider variety of situations, including economic and diplomatic cases. MMIs that allow for the projection of force faster and in greater depth could still allow states greater coercive options in regular interactions, meaning they should influence the way that states behave even if both states have achieved an MMI.

It is possible that rapid mimicry may erase the speed-based impact of an MMI, meaning that neither side gets an advantage once the first-mover advantage has been countered. Still, MMIs should enable states to issue credible threats in a shorter time frame than was previously possible. Also, a state that has experienced an MMI could have the ability to respond to diplomatic overtures much more quickly than was previously possible. The end result is that the rate of military interactions between countries should increase at the sub-warfare level (i.e., low-level militarized disputes) because of the increase in the opportunities for interaction offered by an MMI.[62]

INTENSITY

Diffusion patterns will also influence the rate and extent of casualties, or war intensity. If slower-spreading MMIs lead to more asymmetrical wars between MMI adopters and nonadopters, those wars are likely to be shorter. Yet the wars themselves will probably feature much higher relative casualty rates suffered by the nonadopter, controlling for war length and engagement size. In contrast, wars fought during periods of rapidly spreading MMIs are likely to be especially intense as states deploy heightened military capabilities against each other and raise the overall level of violence without changing the balance of forces, meaning the war is likely to be both bloodier and longer.[63]

---

[62] Alternatively, MMI adoption could increase the number of choices for an adopting state in a given militarized situation, lengthening the decision-making process.

[63] As described above, innovations are themselves rarely offensive or defensive; the question is whether one side has an asymmetrical advantage. If an innovation tends to make attacks easier if the defender has not adopted, though, versus making defense easier if the other side has not adopted, it could lead to some advantages. This line of thinking shows the importance of organizational plans for using technology, rather than technology in and of itself. Technology only acquires an offensive or defensive character in the hands of military organizations. See Goldman and Andres 1999.

Military Interaction Hypothesis: In general, states that experience an MMI will have more frequent, varied, and intense military interactions with a broader range of states than those that have not experienced an MMI.

ANSWERING OBJECTIONS

There are several potential questions raised by adoption-capacity theory.[64] One clear criticism is that this story only deals with the "supply-side" of the equation, ignoring the issue of why states might be interested in adopting innovations. This is an important related issue, but this book is designed to build on existing work and deal with a distinct question. As noted above, there has been a great deal of prior work, much of it good, on why states are interested in adopting innovations. Adoption-capacity theory begins at the point that diffusion starts, *after* the point at which the international system, or norms or other factors, influence whether or not a state feels compelled to respond to an innovation. It is variation in the adoption requirements for innovations in relation to the capacity of a state to adopt that determines the *content* of state responses. The geostrategic environment can make a country want to respond to an innovation, but state capacity to adopt will be a critical input into the character of that response. That analysis helps a state figure out whether adoption, using external alliances, or shifting to neutrality, for instance, might be the best response. And of course, those choices are not necessarily exclusive.

Critics of the idea that organizational capital matters might protest that rational national leaders should recognize that their military organizations are ossified and adjust national policy accordingly to boost the organizational capital of their military, giving them the capability to adopt necessary innovations. Domestic political constraints, however, often prevent leaders from enforcing large-scale organizational changes on a national military in the short-term. Most national leaders, even in nondemocracies, depend on support from at least one and sometimes more major nonmilitary constituencies in addition to support from the military. Barbara Geddes's work (1999) explaining the restraints on leaders in different types of nondemocracies shows that few leaders have true dictatorial power to quickly enact large-scale changes. The process of building support takes time and bureaucratic political capital. Given that leaders want to stay in power, they are unlikely to push for changes that they think will decrease their chances of political survival, especially if ejection from office may make their personal survival less likely as well (Goemans 2000). So if the short-term risk to the leader from a "nonoptimal" response is lower than if

---

[64] For instance, there are potentially significant questions regarding how power transitions function and the war-escalation process that are excluded for space-related reasons.

the leader pushes for adoption, most leaders will rationally decide not to push for adoption.

Rosen's work (1991) on military innovation also shows that external pressure from politicians is often not enough to cause militaries to change their organizational setup, especially pressure related to war fighting. For example, many link the operational struggles of the U.S. Army in the Vietnam War in part to its failure to reform despite pressure from civilian leaders. In the early 1960s, U.S. president John F. Kennedy directly ordered the army to focus on and train for counterinsurgency warfare. Despite this high-level pressure, the army failed to significantly change, instead opting to cosmetically modify its training programs and define counterinsurgency as a lesser and therefore easier form of conventional warfare (meaning training for conventional war would also prepare the army for counterinsurgency warfare). As a result, the U.S. Army entered the Vietnam War much less able to fight an insurgency than it would have been if external civilian pressure had been enough to cause change (Bacevich 1986).

Another potential critique is that perceptions of innovation effectiveness, rather than adoption capacity, predict the distribution of responses to an innovation. If countries only adopt those innovations that work, does this provide evidence against adoption-capacity theory? This question misconstrues the point of adoption-capacity theory. It is certainly true that perceptions of the effectiveness of an innovation create the framework within which diffusion does or does not occur. For example, strategic bombing is an innovation that did not diffuse in part due to perceptions of ineffectiveness. No theory can explain everything. The point of adoption-capacity theory is that within the range of those innovations that states might be interested in adopting, wanting to adopt an innovation is not enough to predict the distribution of responses or the affect on international politics.

Moreover, if knowledge of innovation effectiveness is all that is necessary to predict diffusion, there should be either a lot more diffusion of innovations than happens empirically—since most MMIs do not spread widely, especially initially—or all actors with the capacity to adopt an innovation should adopt. But instances like the British Army in response to Napoleonic warfare, which had the capacity to adopt mass mobilization but chose not to, show that countries do not automatically mimic successful innovations just because they can (Herrera and Mahnken 2003, 216–17). Also, if perceived effectiveness is the key factor, the pattern of responses to innovations should not change based on the adoption requirements, meaning empirical variation in the number of states that adopt, counter, and/or form alliances shows that automatic adoption of successful innovations does not occur. Instead, it is the relative financial and organizational requirements for adopting an innovation that best predict the distribution of responses and many individual actors.

Finally, in many cases, only a small number of actors either adopt an innovation or have the ability to adopt. These relatively small sample sizes, rather than representing a problem for the theory, help illustrate the difficulties involved in adopting especially complicated innovations. If innovation adoption was easy and many states could quickly adopt major military innovations, even though how wars are fought might change, a given innovation would be unlikely to change the international security environment much in a *relative* sense. The lack of innovation diffusion in some situations, despite clear strategic need, demonstrates adoption-capacity theory at work.

This book represents an attempt to fuse together some important aspects of previous research into a more coherent explanation for the diffusion of military power. Therefore, when assessing alternative arguments, in many cases pieces of them overlap with adoption-capacity theory, although pieces of them diverge as well. Additionally, in some instances I explain what alternatives from the literature might look like even though the literature does not explicitly address the same questions. That is a crucial caveat on this section, but it also means I am trying to build the strongest possible alternatives for comparison. Drawing from these preexisting theories means this section focuses more on attempts to derive orthodox and exclusive versions of alternative explanations than anything else.[65]

Returning to the existing research on the diffusion of military power within the international relations literature, as described above it is possible to derive three competing schools of thought: a more or less neorealist set of arguments concerning the critical role threats and strategic competition play in governing the diffusion process; a norms-based contention that innovation adoption occurs when states try to gain status or legitimize their existence, not as a strategic measure to increase relative power; and a cultural assertion about the importance of cultural similarity in allowing responding states to adopt innovations. As explained above, adoption-capacity theory both builds on and moves beyond this research, but it is essential to discuss some of the distinctions.

Strategic Competition

The broadest version of this realist line of argumentation is that states engage in security maximization in response to military innovations, with geopolitical pressure based on the particular situation of a state determining the particular national response to an innovation. The behavior of military organizations in response to innovation is determined by systemic pressure.[66] Resende-Santos's

---

[65] For example, while interservice competition undoubtedly plays a critical role in the process in some respects, there is not an account of its relationship to the diffusion process writ large. Moreover, the understanding of organizational capital presented here incorporates some elements of interservice rivalry.

[66] Goldman and Eliason (2003b, 8) also make this point.

recent book (2007) on South American militaries in the late nineteenth century maintains that the severity of external threats explains the extent of emulation and that countries attempt to emulate successful innovations in relevant areas of warfare.[67]

The strategic competition alternative is powerful because most instances of military diffusion probably have *some* strategic aspects at their core, whether it is due to perceptions of threats motivating interest groups to promote particular types of changes or global norms causing the interpretation of events to occur in such a way that it makes certain innovations more likely to diffuse than others. Indeed, when applied to the decisions of particular states, adoption-capacity theory similarly contends that the geopolitical position of a state helps predict its behavior—though in a less mechanistic fashion.

Nonetheless, there are several reasons to think the picture offered by the strategic competition perspective is incomplete, especially in the case of Resende-Santos's research (2007, 296, 300), which is the most recent realist work on the topic. First, a strict strategic competition approach excludes the notion that financial and organizational capacity play important roles influencing the response to innovations. As described above and as the subsequent empirical chapters will show, however, factors internal to military organizations play a critical part in determining how militaries approach war and organize themselves to fight. Case studies ranging from British strategy in the Boer War to U.S. naval developments in the 1920s and 1930s to Soviet military innovation in the post–World War II period show the significance of internal military organizational factors in driving the innovation-adoption process (see, for example, Avant 1994; Farrell 1998, 408–9; Rosen 1991; Zisk 1993).

Second, strategic competition can predict potentially which states might be interested in adopting an innovation, but it does not predict which of them will actually succeed in adopting; it can only predict interests, not outcomes, since it cannot explain capacity (for a similar point, see Goldman 2003). For example, Resende-Santos (2007, 9) explicitly excludes whether or not attempts at adoption succeed from his theory. While a neorealist might respond that sufficiently motivated states will always just build the capacity to adopt an innovation, the discussion above shows that sometimes the organizational or financial barriers to change may be sufficiently large that adoption becomes unrealistic. Adoption-capacity theory predicts both the strategy a state will adopt and whether or not attempts at adoption, if they occur, will succeed.

Third, a pure strategic competition approach risks tautology: if the theoretical prediction is that states adopt military innovations when they feel strategic pressure, one could argue that by definition, anytime a military adopts

---

[67] Waltz (1979, 92, 106, 127) suggests that emulation rapidly occurs in response to innovation and that security maximization is only really a question of adoption. Elman (1999), however, disagrees.

an innovation it demonstrates the "capability," broadly defined, to adopt the innovation and the state must have faced a strategic threat, while anytime an innovation is not adopted it demonstrates the *lack* of capability or need, broadly defined, to adopt. Threats do clearly matter and influence the way that actors view the potential risks involved in making defense planning choices. But threats alone are insufficient; questions of the relative benefits or costs of innovations are something that several theories could potentially explain, meaning it is inappropriate to describe all approaches that argue states weigh costs and benefits as "realist" (Zarzecki 2002, 67).

In addition to these arguments, there are other important distinctions between adoption-capacity theory and Resende-Santos's claim. He argues that countries emulate military systems with proven success in warfare, using the South American adoption of the German military model (Resende-Santos 2007, 80–81). But the international response to the *Dreadnought* shows that large-scale responses to a new innovation, both emulation and alternative strategies, can occur even before use in warfare (see chapter 5). Adoption-capacity theory is also focused on something smaller since it concentrates explicitly on the response patterns to the debut of a particularly large and crucial class of military innovations, rather than a situation where an established model has existed for decades. Additionally, while he only looks at whether or not states emulate, adoption-capacity theory predicts the alternative response strategy of states, which could be emulation, but could also be something else. Finally, adoption-capacity theory makes predictions about the systemwide response to an innovation, while Resende-Santos addresses how individual states might respond.

International and Domestic "Norms"

Another perspective on the diffusion of military power maintains that shared norms of appropriate behavior determine the way that states respond in the international environment and that military affairs are no exception (DiMaggio and Powell 1983; Finnemore 1996). One phrase often used to describe this approach to studying military organizations is "sociological institutionalism" (Hall and Taylor 1996).

Dana Eyre and Mark Suchman apply sociological institutionalism to military technology purchases in the developing world, contending that the spread of military technology is based more on perceptions of appropriateness by minor powers than genuine security concerns (Eyre and Suchman 1996, 96; Suchman and Eyre 1992; Wendt and Barnett 1993). Theo Farrell's work (2005) on convergence due to norms, described above, also fits into this category.[68]

---

[68] There are several variants of these arguments, including competitive isomorphism, or mimicry due to competitive pressure, which leads to predictions similar to the realist claims explained above.

Institutional isomorphic approaches often predict mimicry due to status seeking rather than competition.

To some extent the norms alternative is not necessarily competitive with adoption-capacity theory, because adoption-capacity theory is not deterministic on the reasons why states might attempt to purchase specific technologies. Because the locus of the theory involves systems of fighting, or the organization of military forces for warfare, it is entirely possible that weapons transfers, especially, occur at least partially in line with the predictions of this norms research.

While norms may play an important role in determining the rate and extent of diffusion, the norms approach is also incomplete (though to be fair, scholars like Farrell acknowledge that norms are not the only variable that matters). First, even if actors adopt organizational approaches due to the influence of international norms, those norms may exist as heuristics that maximize efficiency for nation-states in the international system. Some research in economics suggests that in "noisy" environments, or environments in which there are multiple conflicting signals, the adoption of goods or services by high-status actors functions as a proxy assessment of the quality of the good. That is to say, nation-states may converge on similar ways of producing military power because they are the best practices, not because they want to model other states to legitimize their existence.

Most important for studying the diffusion of military power, even if a norms perspective can explain the reason actors are interested in choosing particular strategies, it cannot explain implementation success or failure. Either a more fine-grained perspective or an approach where the norms motivating interest in the adoption and implementation of the innovation are relatively identical is necessary. A comparative advantage of adoption-capacity theory is its ability to describe the range of strategic choices in response to innovation, and whether those choices, including adoption, are likely to lead to success or failure for a given state.

"Culture"

Building on research inside and outside international relations, Goldman has shown that the levels of cultural tolerance help explain when states will make serious efforts to adopt innovations from abroad. Political systems that are more tolerant of external ideas are more likely to adopt innovations created abroad, since the public and elites will not view the innovation as an existential threat (Goldman 2006, 70). This is consistent with Michael Fischerkeller's research (1998) on the cultural foundations of military assessments.

It is essential, though, not to totally discount the significance of survival motivations for determining military organizational approaches, as Goldman (2006) also recognizes. While elites may block organizational changes that undermine their power or influence, the incentives to adopt efficient ways of producing military force will be powerful and frequently generate countervailing interest groups. The threat levels facing a country or the initial relative power of interest groups may get filtered through a cultural viewing lens—but they may have independent power as well, influencing the way domestic actors perceive the necessity of change.

Goldman (2007) and Elizabeth Kier (1997) also recognize that culture is not static. The debates over the implementation of a particular military organizational approach could lead to minds being changed and the creation of new cultural norms, especially if the approach is adopted in response to shocks like an unexpected defeat in a war.[69]

Second, arguments often called cultural may also fit into domestic political or organizational approaches, meaning determining causal priority is difficult. For example, while elites want to survive in their positions of authority and might view the adoption of some particular military organizational approaches as threatening, it is difficult to draw the line between calling that sort of rigidity and power seeking a "cultural argument" versus a "domestic politics argument," with the organization of the state or military playing the determinative role in predicting the outcome of attempted military organizational changes.[70]

One limitation with this cultural alternative is that it is only focused on the question of innovation adoption; it does not make predictions about strategic choices in general. The idea that cultural openness may influence the interest states have in adopting innovations is also consistent with adoption-capacity theory, since adoption-capacity theory does not exclude other motivations for adoption.

Finally, the strong cultural perspective also risks tautology in some cases because it is difficult to evaluate variables like openness to cultural diversity without evaluating if the state in question actually adopted the particular innovation being studied.[71] This risks the independent variable (the level of acceptable cultural diversity) being coded on the basis of variation in the dependent variable (whether or not the innovation is adopted). This is clearly problematic; it both artificially creates the appearances of a causal linkage where one may not exist and could lead to attribution bias because it incorrectly identifies the actual source of cultural diversity, failing to independently define that variable.

---

[69] Another possibility is that culture may sometimes serve as a proxy for substantive debates, with cultural language serving as a way of explaining material concerns. The point is not that culture does not matter, just that it is important not to view culture as necessarily static.

[70] Goldman (2006) explicitly tests these arguments against each other in her work on the Ottoman Empire and Meiji Restoration, finding that political culture, in the form of elite beliefs, influences the receptivity of a society to military innovations. Adoption-capacity theory attempts to provide an explanation for issues like why elites' attitudes might be more or less receptive to innovations in a given case.

[71] Goldman's work (2006) explicitly hedges against this possibility.

Adoption-capacity theory can also arguably describe the conditions in which cultural openness is likely to occur, especially with regard to military organizations. Militaries with high levels of experimentation, young organizational ages, and critical tasks less defined by operational methods have higher levels of organizational capital, making it easier for them to be culturally open and adopt innovations. This is a more complete argument explaining innovation capacity because it allows for concrete predictions about when cultural tolerance is likely to occur, along with falsifiable indicators such as the commitment a military has to experimentation.

CASE SELECTION AND NEXT STEPS

This book employs two primary research methods—qualitative cases studies and quantitative analysis—to determine how states respond to new major military innovations, and how those responses affect international politics. The case studies are necessary because, as discussed above, differences in how long innovations remain relevant and the possibility of overlapping innovations in different areas of warfare, like land and sea, make a comprehensive data approach untenable. Additionally, innovations occur in part as an action-reaction process. New innovations displace, or sometimes even exist alongside, the dominant weaponry and strategies of previous periods. This means attempts to measure the diffusion of innovations are often "right censored." We rarely get to see the full distribution of responses to an innovation, over time, since new innovations arise to replace old ones while states may be trying to build the capacity to adopt or develop a long-term response strategy. In other cases, we might get to see the full distribution. But the potential for right censoring makes it difficult to evaluate the diffusion process.

Instead, the unit of analysis is the innovation itself. Process tracing is used for each case to establish the existence of the innovation and evaluate the extent to which diffusion processes governed the specific strategic choices of key states. Quantitative analysis provides a bigger-picture perspective for some of the cases, showing trends only comprehensible when the reactions of all states are aggregated together at the system level. The methodological combination increases confidence in the accuracy of the results by ensuring that both the micro- and macrolevels are consistent with the theory.

Table 2.4 shows a range of military innovations that scholars have claimed "qualify" as major military innovations from 1800 to the present. Since many scholars diverge in their thinking about which cases count, this study uses the universe of major military innovation candidate cases as its range.[72] Even if

---

[72] The debate over which military innovations count is an important issue that deserves recognition. This book can only examine a small number of cases, and there are disagreements over which

TABLE 2.4
Range of Possible Major Military Innovations, 1800–Present

Timing	Revolution
Nineteenth century	Napoleonic revolution/levée en masse
Nineteenth century	Strategic communications/mobility
Nineteenth century	Professional staff personnel and procedures
Nineteenth century	Prussian open-order tactics: Railroads/rifles/telegraph
Early twentieth century	**Battlefleet warfare**
Nineteenth to twentieth centuries	Tactical fires (machine gun and artillery)
Nineteenth to twentieth centuries	Medical
Nineteenth to twentieth centuries	Fortifications (trenches)
Twentieth century	Chemical weapons
Twentieth century	"The modern system"
Twentieth century	Total industrialized war
Twentieth century	Blitzkrieg
Twentieth century	**Carrier warfare**
Twentieth century	Tactical air attack
Twentieth century	Air warfare (bombing)
Twentieth century	Submarine warfare
Twentieth century	**Nuclear weapons**
Twentieth century	Mao Tse-tung's "people's war"
Twentieth century	**Unconventional war/suicide terrorism**
Twentieth century	Microelectronics/genetics engineering
Late twentieth century	Information war/network-centric warfare
Late twentieth century	Fourth-generation warfare

scholars disagree about which cases meet their particular definition, they generally agree that the set of cases under consideration represents the most important military innovations. The existence of the scholarly debate itself also indicates that it makes sense to select cases from the broader set.[73]

The major military innovation cases evaluated in this study are the following: early twentieth-century battlefleet warfare, mid-twentieth-century carrier

---

cases should count. But this book represents a first step toward a more integrated theory of innovation diffusion.

[73] Some of the cases on this list overlap because scholars disagree in the way they categorize an innovation. For example, some scholars break up innovations like Prussian open-order tactics into three technologically defined component innovations: railroads, rifles, and the telegraph. The coding is based on information gathered by the author and Alexander Roland (1997). The bold cases are those covered by this book in depth.

warfare, nuclear weapons, and suicide terrorism. As table 2.5 shows, this selection of cases maximizes variation on the two key independent variables: financial intensity and organizational capital. It also allows for significant time variation, over a period of a century, and cases that focus on both nation-states and nonstate actors.

Carrier warfare is one of the only (strategic bombing may be another) major military innovations requiring high levels of both financial intensity and organizational capital to adopt. Even during the interwar period, when the battleship was still considered the most important element of the modern navy, aircraft carriers had already overtaken the battleship in terms of the cost per unit. Aircraft carriers are as complicated to operate as they are expensive, especially the adoption of carrier task forces, the crux of the carrier warfare innovation. Operating a floating airfield and the ship itself, plus coordinating with support ships, is simply a much harder set of tasks than lining up the big guns of a battleship and firing.[74] In addition to this operational complexity, the critical task of carrier warfare, the massed air strike, replaced the five-hundred-year stranglehold of the big gun over naval warfare. So carrier warfare represents a disruptive innovation requiring enormous levels of organizational capital to implement, alongside its high cost.

Nuclear weapons, like carrier warfare, require high levels of financial intensity for adoption, but in contrast they only require low levels of organizational capital. Nuclear weapons are the only weapon in history where the mere possession of one unit, one nuclear bomb, regardless of anything else, substantially impacts international security, as the unresolved North Korean and Iranian nuclear controversies highlight. There is no other weapon where just threatening to develop a single decades-old version vaults a country onto the priority list for high-ranking U.S. government officials. The uniqueness of nuclear weapons as a symbol of power for coercive purposes in and of themselves, and the potential destruction in the case of use, makes studying the spread of nuclear weapons and the strategic choices made by states in response to nuclear weapons especially interesting. It requires a low level of organizational capital to adopt, despite its organizational complexity, because there is no particular organizational form that is necessary to acquire and deploy nuclear weapons.

The suicide terrorism case is distinctive because it involves an attempt to apply a theory designed to predict the behavior of traditional military organizations to terrorist groups—nonstate actors. Suicide terrorism represents the premier military "innovation" among tactics by terrorist groups (and arguably all nonstate actors) over the last twenty-five years.[75] The chapter focuses on the

---

[74] While battlefleet warfare was not easy to adopt either, carrier warfare was harder.

[75] Depending on how one thinks about guerrilla warfare, and whether current nonstate violent movements are substantially similar to or different from past guerrilla warfare strategies, "modern" guerrilla warfare could also qualify.

TABLE 2.5
How Variation in Adoption Capacity Drives Case Selection

	Required level of financial intensity	*Required level of organizational capital*
Battlefleet warfare (chapter 5)	Medium	Medium
Carrier warfare (chapter 3)	High	High
Nuclear warfare (chapter 4)	High	Low
Suicide terrorism (chapter 6)	Low	High

terrorist group as the unit of analysis, looking at the factors that drive some terrorist groups to adopt suicide terrorism but others to stick with prior tactics. Some revisions are necessary to account for the differences between states and nonstate actors, yet the core theoretical contentions are pertinent. Terrorist groups still face organizational challenges and financial constraints, meaning adoption-capacity theory is applicable.

As referenced above, the adoption of suicide terrorism requires a low level of financial intensity. There is also no real "signature" technology underlying the innovation. Delivering explosive destruction through the necessary death of the transmitter, the suicide bomber, is a different operational paradigm from something like kidnapping or hijacking, however. As an innovation that requires high levels of organizational capital but only low levels of financial intensity to adopt, the suicide terrorism case is a nice contrast to the nuclear weapons one. By showing the portability of the theory beyond simply those MMIs mostly of interest to great powers, this case highlights the utility of adoption-capacity theory for thinking about strategic choices in response to all types of military innovations and perhaps for outside the military realm as well.

Finally, since no cases truly fit in the "low-low" box, further diversification along the key independent variables was conducted by selecting an innovation requiring medium levels of both financial intensity and organizational capital to adopt. The battlefleet warfare case involves the creation of the all-big-gun steel battleship, the culmination of the fifty-year-long operational shift away from the wooden-hulled sail navies that had dominated naval warfare for centuries. Battlefleet warfare also involved skyrocketing procurement costs for the new ships because the fast rate of warship obsolescence in the late nineteenth and early twentieth centuries due to rapid technological changes meant states had to consistently ramp up their naval procurement budgets. Give that preinnovation spending was already high at the unit cost level and the underlying basis of the previous era had also been military rather than civilian, the financial intensity required for battlefleet warfare was medium. Modern battlefleet warfare forced enormous changes in recruiting and training, because the core tasks conducted by the average sailor had changed from climbing the rigging and trimming sails to operating steam boilers and using automated fire control systems for the big guns. Still, battlefleet warfare was also a sustaining innovation because while it required large-scale changes in recruitment, education, and training, the critical task of most navies remained the same: seeking a decisive battle in which it could deliver the maximum possible number of shells on to enemy ships. Thus, the organizational capital required for adoption was medium.

Each chapter proceeds by outlining the essence of the innovation and its emergence in the international system, then applying the theory to predict the strategies states will choose in response, whether those that attempt to adopt the innovation will succeed or fail, and how the overall distribution of choices influenced key aspects of the international security environment. Depending on the available data, the empirical chapters feature different combinations of qualitative and quantitative methods. The suicide terrorism and nuclear weapons chapters include quantitative statistical tests for the diffusion of the innovation, while the battlefleet and carrier warfare cases rely more on process tracing based on primary and secondary source research.

Conclusion

Adoption-capacity theory is a new approach to studying the introduction and spread of major military innovations. It builds on existing research on military innovations and emulation to produce a framework for explaining the way that military innovations spread throughout the international system. It also helps us understand, conceptually, the way that the uneven pace of military innovation diffusion functions as one of the major driving factors underlying shifts in relative power and warfare. The next four chapters empirically test adoption capacity on the cases described above: carrier warfare, nuclear weapons, battlefleet warfare, and suicide terrorism.

Chapter 3

CARRIER WARFARE

As the predominant form of naval power, aircraft carriers are one of the clearest symbols of military strength on earth. Short of the atomic bomb, nothing signifies the power of a great nation like the possession of a fleet of aircraft carriers, able to control the oceans and project power across great distances. It was the transition from battleship to carrier warfare that helped shepherd in the era of U.S. naval supremacy, and the complexities involved in adopting the innovation have ensured U.S. naval superiority ever since. The United States currently possesses an overwhelming advantage in naval power fueled in large part by its dominance in carrier warfare. While the United States presently operates eleven fleet carriers, no other country operates more than one fleet carrier—and only one other country, Great Britain, operates more than one carrier of any kind. Only the U.S. Navy, the Royal Navy, and the World War II era Imperial Japanese Navy have ever adopted the core organizational practices associated with carrier strike operations, the crux of carrier warfare. This is curious given the natural desire of states to protect their own security by building up their military forces with the best available technologies and practices. If the effective operation of aircraft carriers is a critical element of naval power, why haven't more countries adopted the innovation?

This chapter builds on work by Goldman and others on the development of carrier warfare during the interwar period and the early years of World War II, when carrier warfare matured. The chapter focuses on what happened next: the post–World War II gap between the diffusion of aircraft carrier technology and the diffusion of carrier warfare as an operational practice—the carrier strike operations almost simultaneously implemented as the core element of naval operations by the U.S. and Japanese navies after the Battle of Midway. One explanation for the nondiffusion of carrier warfare and the slow diffusion of carriers themselves involves geostrategic requirements: states simply have not needed carriers. Another possibility is that the failure of carrier warfare to diffuse in the post–World War II era lies somewhere between security requirements and the difficulties, financial and organizational, involved in adopting. The high financial and organizational requirements for adoption caused most naval powers to evaluate the costs and benefits of their naval strategy differently than they did in response to the naval innovations of the past, driving a larger proportion of states to drop out of the naval power game, bandwagon, or try to counter U.S. naval supremacy through alternative means.

Given the continuing relevance of carrier warfare in the twenty-first century, this chapter demonstrates why understanding the spread of military power is an especially useful endeavor, not just for international relations theory, but for policy analysts interested in the future of global power. For example, while many U.S. security analysts have worried for years about the possibility of a Chinese push to develop an operational aircraft carrier, if such a move occurs, the high financial and organizational requirements associated with this effort might frustrate Chinese plans and actually enhance U.S. superiority in the short- to medium-term.[1] The financial challenge of building aircraft carriers and the organizational one of operating them are sufficiently high that credible investments by the Chinese Navy[2] might force them to divert funds from their current naval investment program, which focuses on antiship missiles and submarines. The result could weaken the Chinese Navy's antiaccess strategy against the U.S. Navy in the Taiwan Strait in the short- to medium-term, while the organizational barriers to adoption might prove so large that it undercuts the Chinese Navy's overall modernization program in the long run. Alternatively, successful mastery of carrier warfare by the Chinese Navy, while it might take fifteen years, would represent the first serious challenge to U.S. naval supremacy since the early days of World War II. That a challenge has not happened before, even during the height of the cold war, has much to do with the unique challenges involved in adopting carrier warfare.

Carrier Warfare: The Early Years

In the years prior to World War I, every major power established a segment of a military arm dedicated to the airplane and a naval aviation component. By summer 1914, Richard Layman (1979, 241) argues, the navies of the world had already outlined the conceptual range of the possible for the naval airplane.

Despite the theoretically great promise of the airplane, the actual use of the airplane in World War I was limited, especially on the naval side. For the British, Fleet Commander and subsequently First Sea Lord Admiral David Beatty strongly opposed erecting platforms on ships that would block, even temporarily, their gun batteries (Melhorn 1974, 14). This led the British to pursue a ship dedicated to airplanes, resulting in the conversion of the light cruiser HMS *Furious* into an aircraft carrier. The ship played a key role demonstrating the nascent offensive capabilities of naval airpower when its complement of Sopwith Camels attacked German Zeppelin sheds in the Tondern Raid in July 1918. Airpower also played an important part in the April 1918 Zeebrugge

---

[1] This brackets the question of the U.S. response to China's decision.
[2] The formal name of the Chinese Navy is the People's Liberation Army Navy, or PLAN.

Raid, when the Royal Navy attempted to block German submarines and destroyers into Bruges Harbor.[3] But in general, the widespread use of naval aviation remained a promise, rather than a reality, as the war came to a close.

Studying Carrier Warfare

Given the five-hundred-year history of the naval gun determining control of the seas, and subsequent devotion to the battleship by both naval officers and the public at large, it is remarkable that the carrier replaced the battleship as the beating heart of successful naval organizations within a single generation. Also remarkable is that carrier warfare has essentially failed to diffuse since 1945. In contrast to the way that states have traditionally responded to naval innovations, the response to the mature carrier warfare innovation at the end of World War II has mostly been the pursuit of alternative response strategies. Only one country, Great Britain, has adopted both the technological and doctrinal elements of carrier warfare since World War II.[4]

Goldman has studied the development of carrier warfare in the interwar and World War II periods. She explains how the threat environment and organizational factors—both functionalist and sociological—helped the United States and Japan become first movers, while leading to less complete British adoption, the German decision to focus on defensive, land-based naval air operations, and theoretical advances by the Italian Navy that it could not convert into actual naval capabilities (Goldman 2003, 302). This chapter builds on her work to examine the subsequent period and explain how the innovation diffused—or did not really diffuse, in this case—following the clear acceptance of the carrier as "the new capital ship of modern navies" (267). Understanding these developments requires recognizing not just how the strategic environment and norms shaped the desire to implement, but also "the capacity to incorporate the new practice" (301), especially given the availability of alternative strategies.

Carrier Warfare as a Military Innovation

Carrier warfare, as generally defined, is the combined use of fleet aircraft carriers and an array of logistical ships for the purpose of conducting strikes against enemy naval assets and establishing sea control. It uses carriers as mobile airfields, abandoning reliance on naval gunfire as the core of the naval fleet by

---

[3] "Plan for Operation Z.O." and "Remarks by the Sea Lords," cited in Halpern 1972, 460, 466, 472. See also Karau 2003, 455–56.

[4] Even Great Britain eventually abandoned full-length carriers for a small number of vertical and/or short takeoff and landing carriers.

substituting air-launched weapons for the power of the big gun.[5] Compared to battleship warfare, carrier warfare involved new hardware, most notably the carrier itself, and the integration of technological advancements also utilized in other military areas. Carriers also greatly increased the maximum distances possible in naval warfare. The largest World War II–era battleship guns had a maximum range around twenty-five miles. In comparison, carrier warfare featured engagements ranging into the hundreds of miles. This required huge changes in planning for naval engagements, especially surveillance requirements for discovering the location of enemy carriers.

Organizationally, the shift to adopt carrier warfare was similarly large in scope, requiring major changes in battlefield strategy and doctrine. Delivering munitions on target was not just a skill but also the critical task for most navies. The range and caliber of cannons increased over time, sometimes faster and sometimes slower, but the concepts remained the same. In contrast, the method of delivering explosive force in carrier warfare is through munitions launched from airplanes rather than from the ship itself. Defensive strategies also substantially shifted in comparison with the battleship, with most carriers relying on defensive fighter squadrons and antiaircraft fire to prevent its destruction by an enemy air strike.

All of these things required large-scale transformations in recruitment and training at both the officer and enlisted levels. For officers, implementing carrier warfare meant changing the naval promotion system to shift aviation from a sideshow in comparison with serving on a battleship to a central path to promotion. At the enlisted level, it meant broad changes in many of the tasks conducted by enlisted personnel—instead of learning to aim and fire guns, sailors had to learn to launch and recover airplanes (Rosen 1991). Finally, carrier warfare did not just involve operating one ship; it entailed a complex systems integration task, including coordinating air assets and accounting for logistical support ships and other carriers as well.

The Maturation of Carrier Warfare

The Royal Navy leaped ahead of the rest of the world with the debut of the HMS *Furious*, a fully functional aircraft carrier, in 1917, and both the United States and Japan sought to mimic Britain's advances in naval aviation. In 1917, British naval constructor S. V. Goodall gave the United States the plans for the Royal Navy's new aircraft carriers. These plans became the basis of the first U.S.-designed carriers (Hone, Friedman, and Mandeles 1999, 22). The 1921 Semphill

---

[5] There have been many different types of aircraft carriers. For the purposes of this book, this category includes only fleet carriers and the light fleet carriers designed by the United States at the beginning of World War II from converted cruiser hulls. It does not include escort carriers, for example (Bishop and Chant 2004; Chesneau 1984).

Mission, requested by Japan to learn about naval aviation from the Royal Navy, included extensive flight training, air combat training, and technology transfers that helped accelerate Japanese progress (Peattie 2001, 19–21).

By the mid-1930s, however, the Royal Navy had been overtaken in the area of naval aviation. Despite a successful strike by British carrier-launched Swordfish bombers against the Italian Navy at Taranto on November 11–12, 1940, the Royal Navy had a mixed record in World War II. It struggled when matched against the Japanese Navy, as the sinking of the HMS *Prince of Wales* by Japanese naval air assets highlighted. Outside the Taranto attack, an off-the-shelf operation originally planned in the 1930s, the British never really utilized carrier forces for independent strike operations or centered their naval forces on carriers.

There are some parallels between the introduction of naval aviation into World War I and the initial steps toward combined arms warfare on land taken at the end of World War I. In both cases, the British military demonstrated new force employment concepts that turned into major military innovations by World War II. But in both cases, while it was the British military that initiated the incubation period of the innovation, the innovation blossomed under the developmental eye of another military. Germany turned the promise of combined arms warfare into a reality with blitzkrieg, while it was the United States and Japan that turned the promise of carrier warfare into a reality.[6]

After over two decades of development and almost six months of naval war in the Pacific theater of World War II, carrier warfare came of age in the Battle of Midway and its aftermath.[7] Both the planning for the battle—especially decisions concerning the fleet compositions of both sides by Chester Nimitz and Isoroku Yamamoto—and the organizational responses to the battle by the Imperial Japanese Navy and the U.S. Navy, indicate that the Battle of Midway was the real takeoff point for the mature carrier warfare innovation (Fuquea 1997, 710; Hone 1998, 452–53; Office of Naval Intelligence 1947; Wildenberg 2005). For example, in Nimitz's action reports (1942) after Midway, he stated that the priority of long-range air strikes against the Japanese carriers meant carriers had become a decisive element of naval battle and it was necessary to take America's carriers, rather than its battleships, to Midway.[8]

After Midway, carrier warfare continued to mature, with both the Imperial Japanese Navy and the U.S. Navy changing their force structures to feature

---

[6] Thomas Hone, Norman Friedman, and Mark Mandeles (1999, 97–99) make a similar point related to the British failure to take advantage of blitzkrieg.

[7] The British carrier-launched attack at Taranto only involved a single carrier and was not accompanied by doctrinal change. The Japanese Pearl Harbor operation involved multiple carriers but was also not accompanied by a doctrinal change.

[8] In discussing why he did not bring *Task Force One*, America's most modern battleship, to Midway, Nimitz (1942) said, "It was not moved out because of the undesirability of diverting to its screen any units which could add to our long range striking power against the enemy carriers."

task forces centered on the employment of aircraft carriers in combination with screening and logistical ships.[9] On July 14, 1942, the Japanese Navy reorganized its combined fleet around carrier task forces. After initial hesitation on the part of the commander in chief of the U.S. fleet, Admiral Ernest J. King, on June 10, 1943, the U.S. Navy released PAC 10, its new doctrinal statement on carrier operations, which institutionalized carrier task forces (Reynolds 1978, 72). After the war, the Office of the Chief of Naval Operations (1947, 23) described Midway as the point at which "aviation had demonstrated its latent power as the principal offensive element of the new American Navy."[10]

Predicting the Spread and Impact of Carrier Warfare

Required Financial Intensity

Early advocates for airpower, especially in the discussion of bombers versus battleships, believed planes would create enormous cost savings relative to battleships. This is apparent, for example, in post–World War I British memos on the potential benefits of airpower.[11] The era of battleship warfare exemplified by the HMS *Dreadnought* required high expenditures, but carriers were more expensive from the start. For example, the HMS *Nelson* was a frontline British battleship produced after World War I. Completed in 1925 and commissioned in 1927, the *Nelson* cost $36.4 million (Jane's Information Group 1935, 18).[12] In comparison, the USS *Lexington* and *Saratoga*, America's second and third aircraft carriers, also commissioned in 1927, cost over $45 million each (Jane's Information Group 1935, 494). These cost disparities increased substantially after World War II. The HMS *Vanguard*, the last battleship built by the Royal Navy (commissioned in 1946), cost £9 million. The HMS *Eagle*,

[9] To cite only a few accounts, see Kernan 2005; Morison 1949; Parshall and Tully 2005; Prange, Goldstein, and Dillon 1982.

[10] Adam Stulberg (2005, 510) also cites this point as critical for carrier warfare. Another source of evidence for the maturation of carrier warfare comes from civilian compilations of naval power before, during, and after World War II. Beginning in 1898, Englishman Fred T. Jane began publishing *Jane's*, a yearly catalog of global naval capabilities. The catalog includes an entry for each country listing its ships and their capabilities. Tellingly, after World War II the catalog began to put carriers first on its lists, ending a decades-old system of listing battleships first in any entry (Jane's Information Group 1950).

[11] See the First Interim Report of the Committee on National Expenditures, December 14, 1921 (Hattendorf et al. 1993, 890–91).

[12] This number is a conversion from £7,504,055 with a conversion of 4.9 dollars per pound (Officer and Williamson 2006; Williamson 2006). The cost data for this ship and the rest in the chapter are reported directly from yearly compilations of naval arsenals, and presented in the GDP-deflated dollar or pound amount for that year. This is appropriate since it is the cost ratio that matters for the argument.

the first British aircraft carrier fully built after World War II, cost £15.8 million (Jane's Information Group 1955, 11–13). The exact numbers are not as important as the cost ratio, which shows the higher unit cost of carriers. Yet carrier warfare required not only designing and building carriers themselves but also the necessary screening and logistical ships as well as the airplanes to serve on the carrier. This additional cost of planes escalated the cost per carrier far beyond any other naval weapon. The rapid speed of technological change during the beginning of the carrier period also meant high costs. Everything from the ships themselves to the planes required constantly refitting existing pieces and designing new models.

Some of the underlying technologies behind naval aviation had commercial applications, with commercial air flights being only the most obvious reason why private firms might have extra incentives (beyond money gained from defense contracts) to invest in the production of technologies useful for the military. Still, the basic elements of carrier warfare, including the ships, planes, and the ordnance to go on the planes, all have almost purely military application, meaning the financial requirements for adoption are high.

It was also difficult to defray the cost of carriers by investing over a long time period—especially in the interwar and post–World War II periods.[13] As with early twentieth-century battleships, midcentury carriers featured high rates of obsolescence due to the speed of technological developments. Investing a little bit every year would not culminate in a modern and competitive carrier warfare force. Today, decades after the introduction of the innovation, even if a state purchases a prebuilt carrier and avoids developmental costs, the maintenance costs of running an aircraft carrier are still on the order of hundreds of millions of dollars a year.[14] The high costs of maintaining the capability only accentuate the significant financial resources required for adoption. Since the underlying basis of the technology is essentially military and the cost per unit is high, carrier warfare should require a high level of financial intensity to adopt.

Required Organizational Capital

What separates carrier warfare from most other military innovations is that implementation requires a high level of both financial intensity and organizational capital. Unlike nuclear weapons, where organizational challenges exist but the atomic bomb itself can alter power balances, the acquisition of an aircraft carrier is relevant almost entirely as part of a system of fighting designed to maximize its utility. An aircraft carrier by itself, without the organizational

---

[13] Note the important distinction with nuclear weapons (chapter 4), where investing a bit every year can produce something useful due to the relevance of even rudimentary nuclear weapons.

[14] The exact number is less important here than the order of magnitude.

expertise for operations and all the requisite support equipment, is unlikely to matter a great deal in global politics.

Carrier warfare requires a high level of organizational capital for adoption because military organizations have to build and integrate a new type of ship into the navy, and alter the recruitment, education, and training of many personnel. Operating an aircraft carrier constitutes one of the most complicated tasks possible for an organization due to the multiplicity of independent tasks that have to be combined and executed in sequence (Rochlin, La Porte, and Roberts 1987). Carrier operations require integrating traditional naval functions with air functions—a difficult coordination task made even more complicated in a carrier task force where the actions of multiple aircraft carriers and associated ships must be synchronized. As Gene Rochlin, Todd La Porte, and Karlene Roberts (1987) write, "In order to keep this network alive and coordinated, it must be kept connected and integrated horizontally (e.g., across squadrons), vertically (from maintenance and fuel up through operations), and across command structures (battle group–ship–air wing)."

The tasks required to operate an aircraft carrier or a carrier task group also involve extensive tacit knowledge. Classroom knowledge of specific jobs cannot replicate the experience of actually conducting microlevel tasks in sequence with the actions of thousands of other people to successfully operate a carrier. High levels of training and the emphasis on learning through doing allows carriers to function at extremely high levels; carriers become learning institutions where new crew members are mentored by the existing crew and develop new skill sets quickly. The collective nature of tasks on a carrier means that crew members must learn to interact cohesively to prevent operational failure (Weick and Roberts 1993).

The high levels of tacit knowledge required to operate an aircraft carrier and the systems integration tasks only fully learned by operating carriers over a period of years means it is extremely difficult for a navy, no matter the ships or planes it has procured, to adopt carrier warfare. An understanding of the capabilities of the two main first movers, the United States and Japan, helps explain the high requirements for adoption. Both the U.S. Navy and the Imperial Japanese Navy, while not fully adopting carrier warfare until after the Battle of Midway, created organizational conditions favorable to the adoption of carrier warfare during the interwar period.

It is well-known that the U.S. Navy demonstrated a commitment to naval experimentation, especially early in the interwar period, which made it easier for the United States to make the final jump to adopt carrier warfare in the midst of World War II. When he headed the Naval War College in the early 1920s, Admiral William Sims designed a series of tabletop exercises featuring carriers. A generation of naval officers, both aviators and nonaviators, moved through the college and participated in these games, educating themselves on the potential power of carrier aviation (Hone, Friedman, and Mandeles 1999; Till 1996).

These tabletop experiments and changes in the promotion structure due to the efforts of Admiral William Moffett, the head of the Naval Bureau of Aeronautics, were combined with actual naval experiments (beginning with the work of Captain Joseph Reeves expanding the air complement of the USS *Langley* and continuing with the annual naval fleet problems) to work through the question of how carriers would function in future naval conflict scenarios (Goldman 2003, 282; Stulberg 2005). An educational system designed to promote experimenting with new force concepts, a bureaucratic structure that allowed for the promotion of aviators, and opportunities for actual naval exercises all combined to create an environment in which the U.S. Navy was institutionally predisposed to accept the results of Midway and adopt carrier warfare as its core naval strategy.[15]

The Japanese Navy also extensively experimented with aircraft carriers during the interwar period.[16] As with the U.S. Navy and British Navy, Japanese carriers doctrinally functioned in the 1920s and early 1930s as supporting elements, providing reconnaissance, scouting, and protecting Japanese battleships from enemy planes (Peattie 2001, 73). Around 1932, though, the Japanese vision of carrier operations began shifting from a support role to an offensive strike role (though potentially still in support of the battle line as opposed to being the center of operations itself), exemplified by a new committee set up by Japan's Navy General Staff to study naval aviation. Japan's Staff College also started studying carrier aviation tactics, generating a 1936 study on the question of dispersing or grouping carriers for operations (73–75).

Nevertheless, while the United States was forced to rely on uncertain experiments to validate operational concepts for naval aviation, the Sino-Japanese conflict, beginning in 1937, provided a natural experiment for Japanese naval aviation. Japanese pilots gained vital experience during this conflict, and feedback from their successes and failures influenced the next generation of Japanese naval planes and tactics (136–38). For example, in the area of dive-bombing, mid-1930s' experiments and late 1930s' exercises led to the decision to send fighter planes with dive-bombers on attack runs to serve as air cover—a tactic that proved effective in the Pacific War (140).

The U.S. Navy's post–World War II account of naval aviation in the Pacific theater pointed to the high level of uncertainty during the interwar period as driving both a lot of experimentation in general and experimentation with regard to the aircraft carrier in particular (Office of the Chief of Naval Operations 1947, 51). The U.S. and Japanese navies also had relatively young organizational ages. Both had only recently begun serious naval modernization programs prior

---

[15] Hone, Friedman, and Mandeles (1999, 159–60, 189–92) and Stulberg (2005, 517) also discuss how the interaction of these factors made innovation more likely.

[16] Interestingly, Japanese carriers, like the British—and possibly as a result of the Semphill Mission—used closed hangar decks, and the Imperial Japanese Navy did not require planes to have folding wings, severely constricting the possible air complement of Japanese carriers and imposing structural delays on the speed of carrier operations (Peattie 2001, 63–64).

to the First World War. As described in chapter 5, the Japanese Navy developed after the Meiji Restoration in 1868 led to the general modernization of Japan. The United States arguably started its naval modernization program even later. One prominent expert on carrier warfare developments, Geoffrey Till, actually references organizational age in suggesting why the Americans and Japanese became the first movers in carrier warfare. Till (1996, 198) observes that "in contrast to British policy, the Japanese and American, being newer in the business, were willing to make bolder departures."

Both the United States and Japan viewed each other as the most likely possible enemy for a naval war in the interwar period and planned accordingly. The strategic geography of the Pacific meant that both sides would have to project power far beyond their homelands to succeed, especially in comparison to European warfare. The limited ranges of shore-based aircraft in the 1920s and early 1930s (a situation that changed in the late 1930s when the pace of advances in aircraft technology accelerated), and a recognition that air power would play an important role in the next war, even if neither side knew exactly how, led to a commitment to pursue the development of carriers and carrier aviation. Both sides viewed their critical tasks for naval warfare broadly, encouraging the development of carriers and carrier aviation.[17]

Using the characteristics of the first movers and the general organizational requirements of the innovation as a guide, the required level of organizational capital for adopting carrier warfare is high.

Carrier Warfare Diffusion Predictions

Based on the general adoption requirement of high levels of financial intensity and organizational capital, the basic diffusion prediction is that few states, if any, should be able to adopt carrier warfare. The noncumulative nature of the financial investments combined with the extensive tacit knowledge necessary for effective operations means the spread of knowledge about the innovation should not make adoption less demanding over time. This should make the adoption of carrier warfare less plausible for broad range of states, even over a long time frame. Adoption-capacity theory predicts that the innovation will essentially not diffuse apart from a rising great power that wishes to adopt, and even that will take a great deal of time and energy.

Other Strategic Responses to Carrier Warfare

Another potential response after the introduction of the innovation involves bandwagoning with the first mover or balancing with potential adopters against

---

[17] Smaller escort carriers might optimally have different missions, like protecting convoys or antisubmarine warfare.

the first mover. The high organizational capital requirements for adopting carrier warfare should make external balancing through alliances more likely than adoption. With other innovations, forming alliances with adopters makes the transfer of relevant technical knowledge more probable, meaning a state in an alliance has a stronger chance of later adopting itself. With carriers, the high levels of tacit knowledge necessary for effective operations cannot be acquired quickly through an alliance. Partial adoption is also a possibility. Given that carriers also serve as a symbol of naval power, some states might acquire smaller fleet carriers or non-fleet carriers but not adopt the organizational practices associated with carrier warfare.

The difficulties involved in adopting carrier warfare mean the number of alliance candidates for potential bandwagoners and balancers is also small, so states that wish to maintain their geopolitical influence but cannot adopt are also likely to adopt alternative military strategies to counter the innovation.

Carrier Warfare Impact Predictions

The adoption requirements should also produce a large effect on the international security environment. The first-mover advantage from carrier warfare should be large, and lock in a relative power disparity between adopters and nonadopters, especially given the predicted high tacit knowledge requirements for adoption. Given the high levels of resource mobilization and organizational change required for adoption, carrier warfare should be a disruptive innovation, threatening preexisting naval powers and reordering power balances.

Late Adopters Hypothesis: The cost per unit and organizational requirements of carrier warfare should not decline by much over time, meaning it will not be much easier for late adopters to adopt carrier warfare than the first mover.

Major Power Dropout Hypothesis: Major powers that fail to adopt carrier warfare will gradually cease to be major powers in the area of the innovation.

The theory also makes a number of predictions with regard to the relationship between innovation diffusion and war that should apply in the carrier warfare case. One of the primary differences between carrier and battleship warfare is the range involved. The USS *Washington*, a North Carolina–class U.S. battleship laid down in 1938 and commissioned in May 1941, possessed nine sixteen-inch gun barrels fired from three turrets; it was the apex of U.S. naval gun development prior to World War II (Chesneau 1980, 97).[18] The armor-piercing range of the sixteen-inch gun barrels was 21 miles; the bombardment

---

[18] There were two turrets at the bow and one at the stern.

range was 22.8 miles. In contrast, U.S. naval planes at the outset of World War II, inferior to their Japanese counterparts in many areas, possessed minimum ranges in the hundreds of miles. The first version of the Grumman F4F Wildcat naval fighter plane had a range of about 770 miles, while a 1943 information sheet from the U.S. Bureau of Aeronautics put the range of the F4F's successor, the FM1, at 830 miles without an external fuel tank and another 1,000 miles with an external fuel tank (Bureau of Aeronautics 1943; *Naval Aviation News* 1971). The range of fighter aircraft surpasses that of battleship guns by almost 3,000 percent.[19] So states that adopt carrier warfare, even holding other things constant, should be involved in more militarized disputes, and more intense ones, with a wider geographic set of states than nonadopters.

THE DIFFUSION OF CARRIER WARFARE

The failure of carrier warfare to diffuse far beyond the initial adopters—though several states have purchased or built individual aircraft carriers—precludes quantitative statistical tests. Additionally, most of the activity involving aircraft carriers in warfare occurred during World War II. The most frequently used international conflict data sets—the MIDs and COW data sets—incompletely reflect the multifaceted nature of a conflict like World War II and are generally considered suboptimal guides to the period. This means that testing the carrier warfare hypotheses, like in the battlefleet warfare chapter, requires more of an emphasis on process tracing.

Preinnovation Naval Powers

After the conclusion of the first World War, and the destruction of the German and Austro-Hungarian navies, along with the rapid decline of Russia into civil strife, Great Britain, the United States, and Japan were the three most prominent global naval powers. France and Italy also possessed sizable fleets. The Washington Naval Conference in 1922 attempted to formally regulate the development of naval warships by the principal powers. The Five-Power Naval Limitation Treaty that resulted from the negotiations established a tonnage ratio for capital ships (battleships and battle cruisers) and aircraft carriers of 5/5/3/1.67/1.67 for the U.S., British, Japanese, French, and Italian navies,

---

[19] This result is not limited to the F4F and FM-1. Data on the TBF *Avenger*, the U.S. Navy's World War II era torpedo bomber, show similar ranges despite having to carry a heavier load than the fighter planes (a torpedo).

respectively.[20] Table 3.1 shows an estimate of the naval strength of the five post–World War I naval powers, plus a resurgent Germany, in 1932 and 1939.[21]

Diffusion Results

Figure 3.1 compares the spread of the core technology, aircraft carriers, to the spread of carrier warfare, which turned the aircraft carrier into the capital ship of the modern navy. The enormous disparity between the spread of the technology and the spread of the innovation highlights the difference between technological change and changes in force employment.

If carrier warfare is the ability to conduct strike operations with carrier-centric task forces and supporting ships, after its introduction the innovation has arguably never diffused beyond Great Britain at the tail end of World War II. No country besides the United States, Japan, Great Britain, and possibly France has ever had enough operational carriers at once to form a carrier-centric fleet. India, in addition to these countries, also once used a carrier for a strike operation with fixed-wing aircraft. So even an expansive definition of adoption still reveals a constrained diffusion pattern. Following World War II, the U.S. Navy again heavily relied on carriers in the Korean War. Their demonstrated utility ensured that carrier modernization became the top U.S. naval priority. The carrier continues to function as the centerpiece of the U.S. Navy (Gardiner 1993, 19–20).

From the 1950s through the mid-1970s, France operated three aircraft carriers—a number that dipped to two in 1974 with the decommissioning of the British-constructed *Arromanches*, but rose again to three in 1994 with the commissioning of the *Charles de Gaulle*, France's only nuclear-powered aircraft carrier, before declining to just the *Charles de Gaulle* at present.[22] While the Soviet/Russian Navy deployed Kiev-class carriers in the 1970s and early 1980s (*Kiev*, *Minsk*, *Novorossiysk*, and *Admiral Gorshkov*), and the *Admiral Kuznetsov* in the 1990s, it never fully adopted carrier warfare.[23]

---

[20] The tonnage allocations for capital ships were 535,000 for the United States and Great Britain, 315,000 for Japan, and 175,000 for France and Italy. The aircraft carrier tonnage allocations were 135,000 for the United States and Great Britain, 81,000 for Japan, and 60,000 for France and Italy. These allocations were capped at the maximums, meaning the ratios presume maximum investments by all states. Of course, a state could choose to invest less, but it would not influence the possible investments of other states.

[21] Data taken from Roskill 1968, 575–78.

[22] France's three carriers in the 1950s consisted of carriers purchased from the United States and Great Britain. The indigenous French Clemenceau-class carriers in the late 1950s were the first instance of aircraft carrier construction outside the United States and Great Britain since World War II (Chesneau 1984, 68–69). The *Jeanne D'Arc* (commissioned January 30, 1964) only operates helicopters and is officially tasked to an antisubmarine warfare mission rather than a strike mission, so it does not fit the technical definition of a fleet carrier (73).

[23] The Soviet case is explained in more detail below.

TABLE 3.1
Interwar Period Naval Strength, 1932 and 1939

	United States	Great Britain	Japan	France	Italy	Germany
			January 1932			
Battleships	12	15	10	9	4	4
Battle cruisers	3					
Cruisers	52	19	27	19	17	6
Aircraft carriers	6	3	3	1		
Seaplane carriers	2		1			
Destroyers	150	251	110	74	86	16
Torpedo boats				4	33	10
Submarines	52	81	67	65	46	
			January 1939			
Battleships	12	15	9	5	4	2
Battle cruisers	3			1		2
Pocket battleships						3
Cruisers	62	32	39	18	21	6
Aircraft carriers	7	5	5	1		
Seaplane carriers	2		3	1		
Destroyers	159	209	84	58	48	17
Torpedo boats	11		38	13	69	16
Submarines	54	87	58	76	104	57

Over the last thirty years, about one new state every ten years has built or acquired an aircraft carrier, though rarely a fleet carrier, and none have adopted the full innovation. At present nine states operate aircraft carriers, with the Chinese Navy a potential adopter given its acquisition of the *Varyag*, a never-completed Soviet carrier of the Admiral Kuznetsov class. Table 3.2 shows the states that currently possess carriers along with the type.

The diffusion of U.S. naval weapons, beginning with the construction of escort carriers for Great Britain during World War II, also provides evidence in favor of the bandwagoning hypothesis. With the destruction of the Imperial Japanese Navy and the growing predominance of the U.S. Navy in the aftermath of World War II, most countries attempted to cooperate with the U.S. Navy rather than pay the cost of an independent naval posture. With the exception of the Soviet Union and China during the cold war, and Russia and China today, nearly every country with an active navy either conducts joint operations with the United States or uses some naval equipment designed by the United States.

Number of States

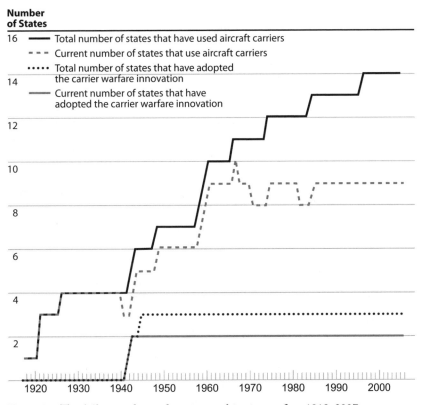

16 —— Total number of states that have used aircraft carriers

‑ ‑ ‑ Current number of states that use aircraft carriers

••••• Total number of states that have adopted
the carrier warfare innovation

—— Current number of states that have
adopted the carrier warfare innovation

Figure 3.1. The diffusion of aircraft carriers and carrier warfare, 1918–2007.

After a military innovation debuts, as Elman and others have explained, one potential response on the part of interested states involves bandwagoning with the first mover or balancing with potential adopters against the first mover. In this case, the high organizational capital requirements for adopting carrier warfare made external balancing through alliances more plausible for most countries than adoption of the organizational practices and the necessary technologies—even given that, as with other military innovations, adopters adapt the organizational practices associated with an innovation to fit their particular context.[24]

The striking shift in the adoption pattern for the carrier warfare case, in comparison with past innovations, highlights these capacity constraints. Following the commissioning of the HMS *Dreadnought* in 1906, almost half of the possible adopters had purchased or built all-big-gun battleships by the end of 1911, and several adopted the organizational practices associated with battlefleet

---

[24] Goldman (2003) makes this claim in the carrier case as well.

TABLE 3.2
Current Global Distribution of Aircraft Carriers

Country	Number of carriers	Air wing type	Propulsion
Brazil	1	Catapult/arrested landing	Conventional
France	1	Catapult/arrested landing	Nuclear
India	1	VSTOL	Conventional
Italy	1	VSTOL	Conventional
Russia	1	VSTOL/arrested landing	Conventional
Spain	1	VSTOL	Conventional
Thailand	1	VSTOL	Conventional
United Kingdom	2	VSTOL	Conventional
United States	11[a]	Catapult/arrested landing	Conventional/ nuclear

Notes: VSTOL stands for vertical/short takeoff and landing, meaning these carriers can only launch helicopters or planes with vertical thrusters that allow for shorter takeoff ramps than conventional or catapulted takeoffs. This carrier design mostly emerged in the 1960s with the development of the Harrier, a plane with vertical thrusters for takeoff (Gardiner 1993, 17). VSTOL carriers tend to function as antisubmarine warfare platforms or assist amphibious landing operations.

[a] This number assumes official information available through the Office of Chief of Naval Operations, U.S. Navy, available at http://www.navy.mil/navydata/ships/carriers/cv-list.asp. Given continuing developments, this total could very well change.

warfare (see chapter 5). In the carrier case, smaller naval powers immediately recognized the importance of carriers to the future of naval power after World War II. In the words of one prominent naval analyst, medium naval powers "had ordered battleships and large cruisers during the heyday of the dreadnought, and it was almost natural that, as the aircraft carrier became the new capital ship, their gaze would turn in that direction (Gardiner 1993, 33–34).[25]

Despite these initial inclinations, though, while most navies still thought about naval power in the terms framed by Alfred Thayer Mahan and the need for sea control, few states ever had a large enough carrier force to actually implement carrier warfare (Till 2005, 326). The subsequent response pattern by medium powers illustrates the unique adoption requirements of carrier warfare.

After the initial diffusion of carriers in the post–World War II era, the total number of countries operating carriers stalled and began declining. With carriers, the high levels of tacit knowledge necessary for effective operations could not (and still cannot) easily be acquired quickly through an alliance, meaning the distributional possibilities through carrier operations existed more on the technical side—the acquisition of physical carrier hulls—than on the operational

[25] For example, in South America both Brazil and Argentina purchased old aircraft carriers before 1960 and configured them for offensive strike operations.

side. The especially high financial and organizational costs of developing carrier warfare meant states that would ordinarily be late adopters felt pressure to make faster strategic decisions in favor of alliances, in addition to any procurement path.

U.S. allies like Canada and the Netherlands that operated aircraft carriers after World War II abandoned them, and bandwagoned even more with the United States (and also Great Britain). Canada, the Netherlands, Australia, and others wished to maintain independent naval strike forces, but the growing costs and complexity of modern carrier warfare overwhelmed their defense budgets. Smaller Western naval powers decided to "abandon carrier-based strike warfare" (Gardiner 1993, 33), and instead bandwagoned with the U.S. Navy, shifting toward a greater focus on antisubmarine warfare and protecting convoys in the case of a war.[26] This occurred not due to changing security perceptions but rather because of the tremendous costs and complexities involved in operating carriers (Gardiner 1993, 34). As Norman Friedman (1986, 88) writes:

> Carrier aviation flowered on a world scale after 1945 because the minimum, in the form of a *Colossus* class carrier, was within the means of many navies. In 1945, it was already obvious that a *Colossus* was far too small and far too slow to be an efficient carrier. That was not the point. It was *a* carrier. As jet aircraft developed throughout the 'fifties, the price of a minimum modern capability rose; navies had to choose between retaining obsolete aircraft and abandoning carrier airpower altogether. India, for example, chose to retain the Sea Hawk fighter, designed to a 1946 British specification. Australia, Canada, and the Netherlands ultimately gave up on carrier aviation.

In the European case, by the 1960s and 1970s, as Till (2005, 312) explains, "domestic social and economic constraints, which actually had nothing to do with East-West relations, maritime or otherwise," drove European naval downsizing. In particular, the increasing price of even staying in the same technological ballpark as the U.S. Navy, given the pace of technological change and baseline price of the technology, forced bandwagoning on the part of all the Western European navies except France, which chose to retain an independent naval structure. The difficulties involved in adopting carrier warfare also meant that the number of alliance candidates for those states interested in balancing or bandwagoning was small. Therefore, European navies refocused on coastal defense, antisubmarine warfare, and specialized sea lines of communication protection (324). The Dutch Navy bandwagoned late in World War II when it abandoned any pretense at a global mission, and instead concentrated on smaller numbers of higher-quality ships that could assist the U.S.

---

[26] Submarines themselves played a critical role in projecting naval power during the cold war and are another case of innovation. For an excellent piece on this topic, see Cote 2003.

and British navies. In "Mainbrace," a North Atlantic Treaty Organization (NATO) naval exercise in 1952, due to their limitations British carriers were confined to a defense and antisubmarine warfare role supported by Dutch fighters—demonstrating the way Britain had bandwagoned with the United States and the Dutch with the British (316, 328).

This is not to say that the security environment did not play a role; it certainly mattered. While the international security environment after World War II differed from previous periods, however, even minor powers maintained large armies and defense expenditures in other areas.[27] Also, despite the bipolar security environment, the initial move by minor naval powers was toward adoption. They did not at first decide to purely bandwagon. Moreover, even after they stopped operating carriers, these navies maintained naval spending, just in an area complementary to the U.S. Navy. For instance, in the 1960s, the NATO Planning Board for Ocean Shipping cited "European promises to augment US efforts with 600 ships" (325). If the costs of operating a carrier force had been lower—comparable to the key technologies in other periods of naval warfare—minor naval powers would have been much more likely to maintain independent forces. The high costs and organizational complexity of operating carriers caused these navies to change how they evaluated the costs and benefits of an independent naval posture, shifting their preferences.

Other states interested in pursuing naval capabilities, but lacking the financial and organizational capabilities to adopt or the willingness to ally with an adopter, have chosen to develop alternative naval capabilities, especially submarines and antiship missiles.[28] They have attempted to counter the innovation rather than emulate. Smaller navies purchase and build an array of capabilities designed to raise the risk of damage to an aircraft carrier entering a particular region to a sufficiently high level that it deters the deployment in the first place.

The dominant offsetting technologies for navies attempting to deal with carriers have been submarines with advanced torpedoes and updated antiship missiles produced in large quantities so they can overcome defensive systems on carriers, regardless of system effectiveness. While nine countries operate a total of about twenty operational aircraft carriers, in 2003 about forty countries possessed a total of about three hundred submarines (McKitrick 2003). Submarines have a much lower unit cost than carriers. For example, in 1935, the U.S. carriers USS *Lexington* and *Saratoga* cost over $45 million each, while Porpoise-class submarines commissioned in the mid-1930s cost about $2.4 million each (Jane's Information Group 1935, 509). Even expensive submarines cost much less, relatively, than carriers. The

[27] For example, many, like Australia, invested in nuclear weapons programs even though they ended up deciding not to build a nuclear bomb.

[28] The Soviet Union emphasized land-based naval aircraft as well, but the submarine and antiship missile became more globally available counters.

nuclear-powered Lafayette-class submarines of the U.S. Navy, produced around the same period as the first nuclear-powered aircraft carrier, the USS *Enterprise*, cost about $109.5 million each. In contrast, the *Enterprise* cost about $444 million (Jane's Information Group 1977, 345, 358).

Submarines also offer a relatively weaker naval power a potential opportunity to sink carriers through torpedo attacks. The experience of the Argentine aircraft carrier *Veinticinco de Mayo* during the Falklands War with Great Britain in 1982, where the British submarine threat forced the carrier into port for the duration of the war, shows how advanced submarines can potentially counter carriers (Fieldhouse and Taoka 1989, 40).[29]

The modern antiship missile represents another possible countering strategy against carriers. The Soviet Navy, facing U.S. superiority in carriers, pioneered the development of antiship missiles. As Richard Fieldhouse and Shunji Taoka (22) write, "The adoption of anti-ship missiles by the Soviet Navy was a revolutionary event in naval history that can perhaps be compared to the introduction of explosive shells to naval artillery, which was also done by the Russian Navy in the nineteenth century." Antiship missiles have a low unit cost, helping minor powers compensate for their comparative weakness in surface ships. Egyptian and Argentine antiship missile success against Israel and Great Britain, respectively, illustrate how these missiles can threaten superior surface fleets (Kojukharov 1997).[30] In 1967, the Egyptian Navy sunk the Israeli destroyer *Eilat* with a Soviet–built Styx missile fired by a Soviet-supplied Komar-class patrol boat, demonstrating the potential effectiveness of antiship missiles against Western technology (Fieldhouse and Taoka 1989).

Despite improvements in submarine and antiship missile capabilities, carrier task forces have also improved their defenses over time with increasingly layered systems.[31] When they are deployed in force with the appropriate escorts, sinking an aircraft carrier is beyond the capabilities of most global navies (Thompson 2001).

The investment pattern of the Chinese Navy has traditionally exemplified the countering strategy described above. The Chinese Navy has invested heavily in submarines and missile technologies designed to counter U.S. carrier capabilities rather than challenging them with Chinese carriers (O'Rourke 2005). The Chinese Navy has attempted to compensate for its inferiority in advanced naval platforms like carriers by focusing indigenous modernization on advanced missile and aerospace components, and purchasing anticarrier platforms like Russian Kilo-class submarines and Sovremmeny-class destroyers,

[29] Ironically, *Veinticinco de Mayo* was the Colossus-class World War II era British aircraft carrier HMS *Venerable* sold to the Netherlands in 1948 and then to Argentina.

[30] During the Falklands War, an Argentine-fired Exocet missile sunk the HMS *Sheffield*, a British destroyer.

[31] These systems range from support ships like AEGIS cruisers and guided missile destroyers to last-resort weapons like the Phalanx Close-In-Weapons-System.

equipped with the advanced SS-N-22 "Sunburn" antiship missile (Crane et al. 2005, 185–87). Recently translated Chinese writings on naval strategy discuss using multifaceted attacks, including operations to overwhelm the sensor capabilities of a carrier battle group followed by waves of antiship missiles and torpedoes, and timing attacks for periods of greater vulnerability, such as during refueling operations. By first targeting the support ships protecting a carrier, the Chinese Navy hopes to decrease the defensive shield and improve the odds of a successful strike against the carrier itself (Mulvenon 2006, 60–67, 73).

The Impact of Carrier Warfare

The adoption of carrier task forces substantially boosted the relative naval power of the United States and Japan by providing a new operational rubric for major naval combat. Centering naval operations on carrier task forces supported by destroyer screens and other logistical ships shifted the carrier from a weapon that could take the tactical offensive to one that was self-sufficient on the ocean for weeks and could execute strategic offensives over great distance.

America's global naval leadership was truly cemented when, following U.S. adoption of carrier warfare techniques, the British Navy reengaged in the Pacific theater in 1945. After its 1941 and 1942 defeats, it found itself as the student rather than the teacher. A report on the British naval experience in the Pacific theater in 1945, written shortly afterward, shows newfound respect for U.S. naval prowess and a recognition that the British had to adopt U.S. carrier tactics (Report of Experiences of the British Pacific Fleet, January–August 1945, Hattendorf et al. 1993, 867).

World War II demonstrated that sea control, the goal of naval powers for centuries, now depended in large part on control of the air (Office of the Chief of Naval Operations 1947, 43). The increasing range of naval aircraft, and the growing size and accuracy of torpedoes and bombs, meant that few targets could escape the reach of a carrier task force. The screening and supply ships included in the task force also made it capable of operating at great distances for long periods of time without returning to land, giving it an enormous operational advantage. These capabilities have dissuaded most potential competitors from even attempting to challenge U.S. surface warfare superiority. America's industrial capacity and operational expertise, which left it with 101 active aircraft carriers in August 1945 at the end of World War II, created a first-mover advantage so large that no state since has attempted to seriously compete with the United States for surface sea control (Chesneau 1984, 41).

It is ironic to describe the Imperial Japanese Navy as gaining a first-mover advantage from the concentration of its remaining carriers in task forces after the Battle of Midway given the elimination of the Japanese surface fleet by the end of the war.[32] The carrier warfare case features a first mover whose military capabilities in the area of the innovation were utterly destroyed just a few years after the innovation matured (Hone, Friedman, and Mandeles 1999, 142, 201–5).

Asymmetrical possession of aircraft carriers, even without full adoption of carrier warfare, has also provided substantial boosts for regional navies. For example, India acquired the light carrier HMS *Hercules* from Great Britain in 1957 and commissioned it as the INS *Vikrant* in 1961. Despite its small size, its status as the only South Asian aircraft carrier provided India with a large increase in naval combat capacity. In its war with Pakistan in 1971, the deployment of the *Vikrant* required Pakistan to deploy a large proportion of its naval assets, like the PNS *Ghazi*, Pakistan's U.S.-made submarine, against the carrier. With Pakistan lacking an effective counter, the *Vikrant* sunk several Pakistani ships and launched strike operations against coastal targets (Pike 2006; Tellis 1985).

The first-mover advantages of carrier warfare are even more starkly apparent when compared with late adopters. Due to the high degree of tacit knowledge necessary to operate aircraft carriers, latecomer advantages appear quite limited. For example, when the Soviet Union built its *Kiev*-class carriers, the first Soviet carriers capable of launching vertical and/or short takeoff and landing aircraft (unlike the Moskva-class helicopter carriers), and began plans for the Kuznetsov-class carriers that would feature conventional takeoffs and landings of fixed-wing aircraft, the U.S. Office of the Chief of Naval Operations (1985, 26) estimated that it would take the Soviet Union at least ten years to master basic carrier operations after the first ship became operational. Given that the *Admiral Kuznetsov* only became operational in 1995 after a twelve-year development program, had the Soviet Union continued to exist, it would have only reached basic proficiency in carrier operations in 2005.

The evidence also supports the major power dropout hypothesis. While Great Britain and France are still defined as major powers according to the Correlates of War Project (2007), their relative naval power has significantly declined since World War II. France currently operates one aircraft carrier, though it is nuclear powered, and Great Britain operates two relatively small aircraft carriers. Comparing the size and airplane capacity of the British Invincible-class carriers (designed and produced from the mid-1970s to the early 1980s), the last completed British carriers, to the U.S. carriers designed during the same period, the early Nimitz-class carriers, highlights the differences between the two. Ships in the original Invincible-class design could displace 16,000 tons

---

[32] Japan did have carriers in reserve after Midway and still had numerical equality in carriers with the U.S. Navy at the end of 1942 (Evans and Peattie 1997, 490–95; Peattie 2001, 200).

and carry 14 aircraft with a crew of 900. In comparison, ships in the early Nimitz-class design would displace 81,600 tons and carry 89 aircraft with a complement of 6,400. Even the U.S. Tarawa-class amphibious assault carriers built during the same period, which only carry helicopters, are bigger than the British carriers. The Tarawa-class design called for ships displacing 39,300 tons and carrying 28 helicopters with a complement (excluding troops on board for an amphibious assault) of 902 (Chesneau 1984, 148, 280–83). Hence, while France and Great Britain remain major powers according to the COW Project, in the area of naval warfare their fleets are so inferior to the United States as to make them nonmajor in this area.

The lack of accurate comparative quantitative measures for naval deployments in the post–World War II era, combined with uncertainty about the accuracy of the MID and COW data for World War II, makes it difficult to quantify the impact of carrier warfare on militarized disputes, the geographic scope of those conflicts, and conflict intensity. Yet U.S. naval deployment data from 1946 to 1990 can help shed some light on the uses of carrier task forces following the implementation of the innovation by the United States. Researchers at the Naval Historical Center in Washington, DC, have identified 207 instances of naval (Navy and Marines) crisis deployments starting in periods of relative calm from 1946 to 1990. Of those 207 cases, carrier deployments occurred in 140, or about 68 percent (Siegel 1991). Nearly all, excluding some in Central America, involve overseas deployments over fairly large distances. So the high pace of carrier deployments and the distances involved seem to provide some initial evidence confirming the effect of carriers on military interactions.

COUNTRY-SPECIFIC RESPONSES TO CARRIER WARFARE

Exploring the naval decision making of several countries in response to carrier warfare after World War II will show how adoption-capacity theory explains the strategic decision making of three potential adopters: Great Britain, the former naval leader; the Soviet Union, the cold war–era great power opponent of the United States; and Italy, a medium-sized naval power throughout the period.

Great Britain

How did Great Britain, the state that exited World War I with the greatest understanding of carrier warfare, come to be a follower by the end of World War II?[33] Goldman (2007) has shown that British force structure plans in the

---

[33] For the early promise of British carriers, see the First Interim Report of the Committee on National Expenditures, December 41, 1921 (Hattendorf et al. 1993, 890–91).

mid- to late 1930s actually looked quite similar to the United States and Japan. But the British decision to house the Fleet Air Arm within the Royal Air Force rather than the navy itself, a lack of commitment to experimentation, and financial constraints all prevented full adoption of carrier warfare prior to 1945. The British maintained a fleet of carriers throughout the war and did successfully utilize them on a number of occasions, most prolifically in the Taranto operation. Nevertheless, failed British naval operations in the early years of the war, like the Japanese carrier strikes in the South China Sea that sunk the HMS *Prince of Wales* and HMS *Repulse*, highlight the problems that the British Navy had in adapting to the changing naval environment (Admiralty Board, April 26, 1937, Tracy 1997, 560–61).[34]

In general, the Royal Navy had to choose between building both battleships and aircraft carriers, which would prevent them from having a critical mass of carriers, or fully embracing carrier warfare. The need for combined operations with the U.S. Navy in the Pacific theater in 1945 drove the British to take the final steps toward adoption. The extensive British investments in carriers and naval aviation in the interwar period and World War II meant that the British already understood how to build and operate aircraft carriers. Close cooperation between U.S. and British naval leaders, beginning with the joint operations in 1945, continued in the postwar period, creating "growing harmony in strategic and operational thought" (Grove and Till 1989, 274).

Interestingly, even after World War II some British admirals contended that the battleship still had a prominent role to play in producing naval power. Yet the presumption about the future of naval warfare had undoubtedly shifted toward advocates of the carrier. An Admiralty memo from July 1945 revealed that battleship advocates now argued for its continued existence due to the fire support it could provide to the aircraft carrier and the need for a "balanced" force where ships would attack their counterparts in the opposing fleet, essentially mirroring the pre–World War II assertions for carrier construction (Admiralty Memorandum to the Cabinet, July 2, 1945, Hattendorf et al. 1993, 949–53).

Financial limitations, which restricted British carrier development during the late 1930s, continued to limit British naval procurement in the post–World War II era. Till (2005, 310) notes that "in 1944–45, the Royal navy had assembled the greatest strike fleet it could produce only to find it regarded as a relatively small task force in the huge US Pacific Fleet." The financial climate forced the repeated slashing of Royal Navy orders for aircraft and a rapid reduction in fleet carriers.

Initial British planning for a postwar navy, discussed in a May 1945 memo by Captain Godfrey French, the deputy director of plans, focused on the continuing need for global power projection and the importance of building

---

[34] See also, among others, Hone, Friedman, and Mandeles 1999, 104–5; Marder, Jacobsen, and Horsfield 1981, 306–8; Roskill 1968, 541–42; Sumida 1993; Till 1979.

different types of ships to help ensure strategic flexibility (Godfrey, May 29, 1945, Hattendorf et al. 1993, 787–92). By 1949, the proposed "Restricted Fleet" for the 1950–58 period was to feature no battleships until substantial modernization of its armaments became possible. Instead, the Royal Navy would maintain five active carriers in peacetime with two additional training carriers, and an expansion capacity to six fleet carriers and four light fleet carriers in a hypothetical 1957 war (Admiralty Board Memorandum, June 1949, Hattendorf et al. 1993, 806). The last British battleship, the HMS *Vanguard*, was commissioned on August 9, 1946.

In the 1950s, British carriers began assuming the traditional British naval role of enforcing sea control and policing the imperial periphery. Till (2005, 310) writes, "A glance at the leading British naval memoirs of the period demonstrates the importance of its traditional East of Suez role to the life and policy of the Royal navy." The Royal Navy engaged in extensive carrier deployments for offensive strike purposes in the cold war, deploying almost as often as the U.S. Navy. The Korean War, the Suez crisis, and the 1961 Iraq-Kuwait crisis all demonstrated the capabilities of the Royal Navy. Britain found its fixed wing carriers increasingly outdated and less able to credibly protect British interests, however, and the Royal Navy had to increasingly bandwagon with the United States (Gardiner 1993, 28–31). This was an organizational shock to a naval organization that had been the most powerful in the world just a generation before. As Friedman (1986, 36) puts it, "The Royal navy had to concede for the first time in its history that its forces no longer sufficed to protect British trade routes. Its future plans would have to assume American participation from the beginning."

British carriers became technologically outdated at the same time that growing financial concerns began overwhelming the British military as a whole, forcing some difficult choices. As early as the mid-1950s, some saw the newly commissioned Audacious-class *Ark Royal* as a "huge and costly white elephant" (Wettern 1982, 107). Budgetary battles escalated throughout the late 1950s and early 1960s as the British undertook a strategic review of their activities overseas in light of changing global circumstances and the British economy.

In response to the growing costs of carriers and declining British capacity to assume those costs, the Royal Navy was forced by Secretary of State for Defense Denis Healey to plan for a future without fixed-wing fleet carriers. A mid-1960s' defense white paper argued that land-based aircraft could substitute for some of the capability of carriers, but more important, the British Navy could no longer afford an active role "East of the Suez" (263–68). As Desmond Wettern (273) observes, "It was clear from Mr. Healey's remarks that the argument against the carriers rested on purely financial rather than on any operational factors." In 1966, the Royal Navy shelved plans for new, large Queen Elizabeth–class fixed-wing carriers, the CVA-01 project, and shifted to building the smaller vertical and/or short takeoff and landing carriers they now operate (Grove and Till 1989, 285).[35] The overwhelming cost of carrier modernization proved too much even for the British Navy (Grove 1987, 272–73). The Royal Navy downsized throughout the 1970s, and the Audacious-class fleet carrier *Ark Royal* became the last conventional fixed-wing carrier in the British fleet before going out of service in 1978. The shift toward vertical and/or short takeoff and landing carriers indicated a strategic move away from maintaining sea control capabilities and toward focusing more on European naval security (Till 2005, 310). Instead, the Royal Navy counted on a combination of its remaining naval assets, nuclear weapons, submarine and antisubmarine warfare expertise, and a close relationship with the United States, especially under the auspices of NATO, to ensure it retained some degree of global naval power projection capabilities (Gardiner 1993, 14).

Soviet Union/Russia

While primarily a land power, the size and scope of czarist, then Soviet, Russia often required significant naval investments. Yet the extreme investments necessary to adopt carriers formed a barrier to acquisition for the early Soviet Navy. In the pre–World War II era, Soviet Navy commissar Vyacheslav Zof argued against the acquisition of carriers specifically due to the adoption requirements, saying, "You speak of aircraft carriers and of the construction of new types of ships . . . at the same time completely ignoring the economic situation of our country and corresponding conditions of our technical means."[36]

Immediately after World War II, Joseph Stalin began a naval modernization program that included developing aircraft carriers. The Soviet Union had received the German carrier *Graf Zeppelin* as part of the postwar settlement. After the ship sank, however, the Soviets lacked a working model, or any operational experience in designing, building, and operating carriers. Given the costs and timelines involved, carrier development took a backseat (Breyer 1970, 190; Yegorova 2005, 164–66). Admiral Nikolai Kuznetsov, the post–World War II leader of the Soviet Navy, wrote in his memoirs that "not excluded either was the construction of aircraft carriers; rather they were only postponed to the last year of the Five-Year Plan. This was explained, I recall, by the complexities of construction of warships of this class and the aircraft designed especially for them."[37]

Instead, with assistance from former German submarine experts, the Soviet Navy initially focused on building submarines and land-based naval assets to

---

[35] See also Hill 1996, 219. Interestingly, the next generation of British aircraft carriers, designed to fly units of F-35s, will likely displace sixty-five thousand tons, about three times the size of the Invincible class. While still much smaller than U.S. carriers, it does reverse the prior British trend toward smaller carriers (Adams 2007).

[36] Quoted in Martin 1970, 48.

[37] Quoted in ibid., 48–49.

counter U.S. naval supremacy rather than challenging it directly through surface ships (Kurth 2005, 265–66). While publicly Nikita Khrushchev and Soviet analysts argued that the nuclear age subsumed the need for carriers (on the grounds that they were vulnerable to submarines and useless in a nuclear conflict), in reality Soviet financial constraints prevented their building carriers. In Khrushchev's memoirs (1974, 31), he wrote, "I felt a nagging desire to have some [aircraft carriers] in our own navy, but we couldn't afford to build them. They were simply beyond our means."[38]

In the late 1950s, as part of a broader effort to seek strategic parity with the United States, the Soviet Union embarked on a blue-water naval expansion program to "compete for supremacy on the high seas" that accelerated after the rise of Leonid Brezhnev (Kurth 2005, 267; Watson and Watson 1986).[39] The Soviet Union decided to more significantly invest in its navy, and move beyond just countering through submarines and antiship weapons. The presence of Admiral Sergei Gorshkov, the "Soviet Mahan," who guided the Soviet Navy from 1956 until 1985, drove Soviet naval modernization. Despite early anti-carrier statements in public, Gorshkov envisioned a "balanced" Soviet surface naval force that included aircraft carriers (Chernyavskii 2005; Fieldhouse and Taoka 1989, 20–22, 40).

First, the Soviet Union constructed two Moskva-class helicopter carriers in the mid-1960s. While these ships were unable to carry anything but a few helicopters, making them useful for antisubmarine warfare purposes, they were the beginning of a long-term Soviet investment in aircraft carriers. From the early 1970s to the early 1980s, the Soviet Navy designed and launched four Kiev-class carriers (*Kiev*, *Minsk*, *Novorossiysk*, and *Admiral Gorshkov*) with the ability to launch fixed-wing aircraft, though only with vertical takeoff, not long conventional ramps or even short "jump" takeoffs (Bagley and LaRocque 1976, 57; Miller 1978).[40] Explained by the Soviets as designed for antisubmarine warfare missions, in reality these ships served as building blocks along the road to fleet carriers. From the 1970s on, the Soviet Navy also began receiving "the 'cream'" of resources within the Soviet military system due to the demands for advanced equipment (Polmar 1983, 76–77, 80).

Soon after the *Admiral Gorshkov* sailed in 1982, the Soviet Navy started work on the Kuznetsov carrier class, the first Soviet carrier class with a full

---

[38] Robert Herrick (1968, 68–69) confirms this interpretation of Khrushchev's viewpoint. See also Chernyavskii 2005, 287–88. Soviet Golf- (starting in 1958) and Hotel-class submarines (starting in 1960) had ballistic missile–launch capabilities, while the Yankee-class submarines deployed beginning in 1968 carried the SS-N-6 nuclear-tipped ballistic missile, the first Soviet submarine-launched intercontinental ballistic missile.

[39] The Cuban missile crisis likely served to bolster the credibility of claims made by Admiral Sergei Gorshkov for budgetary increases.

[40] Sometimes the *Admiral Gorshkov* is considered separately as a Baku-class carrier, but its design characteristics were almost identical to the Kiev class.

conventional deck, aircraft complement, and overall capabilities similar in concept to U.S. carriers. While it was laid down in 1983 and first launched in 1985, the *Kuznetsov*, the only active ship to come out of the program, did not become operational until 1995.[41]

The financial intensity requirements for adoption overwhelmed a Soviet Navy that had to compete with the Soviet Army and other Soviet military institutions for resources throughout the period. While the share of overall Soviet expenditures devoted to the military increased throughout much of the cold war, the pie was far from infinite. It is not a coincidence that the period of greatest Soviet naval modernization occurred after Gorshkov, a five-star admiral with equivalent rank to the heads of the Soviet Army and Strategic Rocket Forces, rose to power in 1956 (Hanks 1981, 122; Hudson 1976b). Still, the Soviets did mobilize billions of dollars to spend on carriers, meaning their investment should have eventually succeeded, based on the experiences of other navies. Yet it did not.

Organizational constraints also limited both the initial Soviet attempts to shift toward carriers and the success of those efforts once they occurred. Western military analysts specifically cited the lack of Soviet naval aviation experience as hindering Soviet carrier development, maintaining that "they will have to experiment and learn slowly the peculiarities of operating high-performance planes from the deck of a ship" (Vernoski 1986, 122). Unfortunately for the Soviet Navy, the hierarchical and secretive nature of the Soviet military apparatus discouraged the type of open experimentation and exercises necessary to really learn the complicated systems integration tasks involved in carrier operations.[42] Bruce and Susan Watson (1986, 292), cold war experts on the Soviet Navy, wrote that the need for centralized control "created many liabilities that can affect the navy's performance. Foremost are the weaknesses inherent in the command and control system. Political considerations clearly have preeminence over independent initiative. Direction comes from Moscow, and authority resides in a dominating naval high command that stifles the initiative of shipboard commanders." Given the operational requirements of carrier warfare, the Soviet system made adoption even more difficult.

Even with a bureaucratic system more conducive to the tacit knowledge and operational tempo necessary to master carrier operations, the adoption-capacity requirements explain why the Soviets struggled. The Soviets would have had to build experience, step by step, even after the deployment of their first conventional takeoff and landing carrier, meaning fielding an effective carrier force

---

[41] The *Varyag* was scrapped short of completion and is now in China. Even after completion, the *Kuznetsov* continued experiencing severe maintenance issues. It reentered port for an overhaul in 1998 and spent several years undergoing repairs.

[42] For a description of Soviet practices in the nuclear arena, see Evangelista 1988.

would take "several times as long" as just building the carriers—itself a difficult task (Breyer 1973, 190). Rochlin, La Porte, and Roberts argued that the organizational characteristics of the Soviet military system would impede its ability to learn how to operate carriers, especially in the short- to medium-term. They wrote that "the Soviet Navy is completely without experience or tradition in large carrier operations. Their internal structure is more rigid and more formal than ours and with far less on-the-job training, especially for enlisted personnel" (Rochlin, La Porte, and Roberts 1987). These predictions proved prescient.

And indeed, naval analysts during the cold war recognized how organizational constraints handicapped the ability of the Soviet Union to field effective, deployable carriers. Even before the deployment of most of the Kiev class, Siegfried Breyer (1973, 192–93) remarked, "Furthermore, the first carrier they built would, undoubtedly, have any number of unforeseen shortcomings which could be rectified only during the construction of later ships. It would take many more years to build the next carriers and make them operational. . . . [I]t would be 13 years at the least before they had even a few operational carriers."

Even if all of its developmental projects had succeeded and the cold war had not ended until the late 1990s, in the mid-1990s the Soviet Navy would have fielded two carriers capable of launching fixed-wing aircraft. While this would have been an incredible engineering and systems integration accomplishment for the Soviet Navy, the resources required to complete the *Kuznetsov* and *Varyag* would have made it difficult to purchase the destroyers, cruisers, and logistical ships necessary to conduct carrier task force operations. And even if somehow the Soviet Navy had found the money to purchase all of the necessary support ships, at the end of the day it would have had one carrier task force, almost no operational experience, and little to no potential for the introduction of more carriers into the fleet for almost a decade.[43] It is striking that even the second-biggest military power in the world did not have the financial resources or organizational capabilities to adopt carrier warfare.

While the Soviet Union failed to adopt carrier warfare, despite its construction of several aircraft carriers, as described above it did attempt to counter U.S. carriers by investing heavily in land-based naval aviation, antiship missiles, and submarines. The Soviet Union was the first mover in the development of modern antiship missiles. The Soviet Union also maintained a larger quantity of submarines than the United States, and while it lagged technologically, produced both ballistic missile and attack submarines designed to ensure its

---

[43] In support of this conclusion, after two decades of Soviet experience with aircraft carriers, analysts with the Stockholm International Peace Research Institute stated that "the US Navy has overwhelming superiority over the Soviet Navy" in the area of aircraft carriers (Fieldhouse and Taoka 1989, 31).

nuclear second-strike deterrent, and provide it with attack capabilities against U.S. carriers (Fieldhouse and Taoka 1989, 52–59; Office of the Chief of Naval Operations 1985).

Italy

The Italian Navy ended World War II in a state of disarray. The World War II story of Italian naval incompetence in the Mediterranean, ranging from the command level down to the technical characteristics of Italy's ships, has been modified in recent years. Research by James Sadkovich (1994, xvii–xviii, 331–32) shows that whether measured by naval vessels sunk, opposition merchant ships sunk, or friendly merchant ships successfully protected, the World War II Italian Navy actually held its own against the Royal Navy. The biggest gap in relative performance came from the Taranto raid in 1940 and subsequent British operations aided by the code-breaking ULTRA program. The naval campaign in the Mediterranean featured some of the clearest examples of ULTRA influencing actual battle outcomes (345).

When Italy switched sides in 1943, the German navy repossessed some of Italy's remaining naval assets. Much of the rest got split among the allied nations after World War II, and Italy was forced to rebuild. Beginning in 1950, the Italian Navy was freed from constraints imposed by the Allies and allowed to begin warship construction again. The Italian Navy's investments focused on destroyers and submarines as Italy's naval strategy increasingly revolved around NATO operations. A law passed by Benito Mussolini in 1923 banning the construction of aircraft carriers stayed in effect throughout the cold war period as part of an effort by the Italian Air Force to block naval aviation (Sondhaus 2004, 290). Yet the Italian Navy has developed a limited carrier force, starting with two Andrea Doria helicopter carriers (*Andrea Doria* and *Caio Duilio*) laid down in the late 1950s. The Italian Navy flagship, the *Giuseppe Garibaldi*, a vertical and/or short takeoff and landing aircraft carrier commissioned in 1985 (though it did not get a complement of Harriers for a few years), is one of the smallest aircraft carriers, as measured by displacement (tons), in the world.[44] While the Italian Navy has built aircraft carriers and purchased planes (mostly Harriers) to operate off of them, the small size of its single carrier, the lack of accompanying screening and logistical ships necessary for strategic operations, and the lack of a doctrine focusing on strike operations means Italy is properly coded as an adopter of some of the technology but not the whole innovation (Jane's Information Group 2006).

Limitations on the financial capacity of the Italian Navy, along with its alliance with the first mover, the U.S. Navy, explain Italy's nonadoption. Italy's

---

[44] The *Chakri Narubet*, the Spanish-constructed carrier operated by the Thai Navy, is smaller than the *Giuseppe Garibaldi*.

response pattern is consistent with the other medium-power navies described in the book. Despite a robust post–World War II economic recovery, as a medium power Italy has never had the financial capacity to modernize at the rate of the larger naval powers, lacking the funds to seriously consider an independent carrier warfare strategy. Additionally, since the early cold war Italy benefited from a formal alliance relationship with the United States. This security blanket allows the Italian Navy to free ride on blue-water operations and long-range strike capabilities—a decision made even easier when the requirements for Mediterranean naval forces declined with the collapse of the Soviet Union. Financial limitations made full adoption impossible for the Italian Navy, and geopolitical developments allowed Italy to bandwagon with the carrier warfare first mover, the United States, at a relatively low cost. This caused Italy to invest more in capabilities to complement the U.S. Navy rather than competitive capabilities like antiship missiles.

AIRCRAFT CARRIERS FOR CHINA?

In 1998, a shell company owned by former Chinese Navy officers purchased the *Varyag*, a partially completed former Soviet aircraft carrier from the Admiral Kuznetsov class. This purchase and subsequent Chinese statements about their desire to build aircraft carriers have triggered great consternation in some corners of the Western press and defense establishment. In 2005, after moving the *Varyag* into dry dock at Dalian, the ship emerged painted Chinese Navy gray. This fact, combined with increasing Chinese interest in airborne-warning helicopters and the carrier-capable version of the SU-33 fighter, has led some analysts to conclude that China is making developing a carrier capability a high priority and that this represents an important strategic challenge to the United States. While the emergence of a China carrier force is certainly not imminent, the Chinese have been gradually increasing their investments in carriers, and the accompanying support ships and necessary planes (Erickson and Wilson 2006, 20–21, 36–37; U.S. Department of Defense 2007, 4, 24). Recent evidence suggests that the Chinese are in the process of building one or two carriers, though their size and purposes is not yet certain (Ross 2009).

What if the Chinese did develop a carrier force? A Chinese aircraft carrier capability would challenge the United States in several areas. For example, Chinese carriers could raise the costs of a potential defense of Taiwan in the case of a Chinese invasion, enable greater Chinese power projection into the South China Sea and beyond, and give China more naval flexibility to pursue its current maritime disputes with Japan. Even though a Chinese carrier force is unlikely to represent a challenge to U.S. naval supremacy in the short-term, there are potentially serious long-term security consequences (Diamond 2006, 55–57; Erickson and Wilson 2006, 27–30; Fisher 2006).

The arguments about adoption capacity outlined in this book suggest that like with the Soviet Union, the path to operating aircraft carriers is likely to be long, slow, and difficult for the Chinese Navy. For example, while the Chinese Air Force has made great strides in recent years, the Chinese unwillingness to allow pilots to train as much as their U.S. counterparts has held them back. The Chinese have been wary about experimenting and practicing too much, lest accidents happen, and they lose planes and pilots. This type of ethic may be detrimental to their ability to effectively employ carriers, since even the U.S. Navy, the most advanced in the world, lost twenty-two planes in 1999 due to carrier accidents. Such an attrition rate is inevitable with active carrier operations, but would be unacceptable in the current Chinese bureaucratic environment (Erickson and Wilson 2006, 25–26).

Given that developing a carrier force is likely to take the Chinese at least as long as it was supposed to have taken the Soviet Union if its carrier program had succeeded, the Chinese development of carriers might actually be a boon for U.S. naval supremacy, especially in the short- to medium-term. If the Chinese divert funds toward carriers, and away from antiship missiles and submarines, this will weaken their anticarrier capabilities—capabilities that U.S. analysts are increasingly regarding as credible threats in the event of a Taiwan contingency (Diamond 2006, 54). Moreover, the Soviet Union spent thirty years trying to develop aircraft carriers, only to find itself several billion dollars poorer and with one partially operational carrier. The systems integration and organizational tasks are sufficiently difficult that it is possible the Chinese would not succeed. In all likelihood, given enough time and money, the Chinese will probably develop fleet carrier capabilities, but whether they can build a capability that seriously competes with the U.S. Navy is an open question. Chinese carrier development efforts are likely to be a lot slower and more uncertain than many analysts have predicted, with potential side benefits for U.S. naval supremacy in the short- to medium-term.[45]

Conclusion

The nondiffusion of carrier warfare, the acknowledged key to naval supremacy in the post–World War II era, is an interesting puzzle for studies of how military power spreads. The immense complexities, both financial and organizational, involved in building and operating aircraft carriers have made it, empirically, one of the most difficult innovations to adopt. Additionally, the

---

[45] For another take on the future of the U.S. Navy, see Hoffman 2008.

overwhelming first-mover advantage gained by the United States after World War II drove U.S. allies to bandwagon, conceding to the United States its carrier-based dominance of the oceans, rather than to adopt themselves. Even Great Britain, the next leading naval power, has never approached more than a fraction of U.S. naval power since the end of World War II.

Those who believe strategic competition is both necessary and sufficient to explain the spread of military power might object that the lack of threats drove the lack of diffusion in the carrier case, instead of variations in adoption capacity. But adoption-capacity theory does not exclude consideration of geostrategic circumstances in predicting state behavior, as chapter 2 makes clear. Moreover, the failed Soviet experiment with aircraft carriers demonstrates a flaw with the pure strategic requirements counterargument. If the Soviet Union was a land power with limited maritime interests, its incredibly costly pursuit of a carrier fleet in the 1970s and 1980s was a colossal waste of money and time that it is difficult to explain. On the other hand, if the Soviet Union did have maritime interests that justified building a carrier fleet, its failure to do so despite massive investments shows how strategic requirements are not enough to cause adoption (Hanks 1981, 122; Hudson 1976a, 96–97). Most important, as discussed above in detail, the high levels of organizational capital required for carrier operations predicted the difficulties that the Soviet Navy had in translating its vision of a carrier fleet into reality. The mixed Soviet experience with aircraft carriers, where a thirty-year investment left it with one barely operational carrier by the mid-1990s, highlights the difficulties involved in operating aircraft carriers even for the largest and best-funded militaries in the international system.

The financial and organizational challenges of carrier warfare drove a large-scale shift in how most states fundamentally designed their naval policies. In the cases of past naval innovations, like battleship warfare, mass emulation followed the introduction of new technologies and tactics. Small and medium-sized navies usually maintained navies that were smaller, but similar in form, to those of the larger powers. The response to carrier warfare is vastly different from that of smaller and medium-sized naval powers to the naval innovations of the past. In the carrier period, the scope of financial and organizational requirements meant having a "small" carrier navy was out of the question for most navies. All but one non-U.S. state with aircraft carriers have only one carrier, and most of those carriers are either tiny or two full developmental generations behind the average U.S. aircraft carrier. Maintaining a navy of smaller size that can do the same things that the U.S. Navy can do is simply not possible in the carrier age. This has forced most navies into either substantial naval bandwagoning with the United States or countering in the form of developing antiship missiles and submarines.

The complicated systems integration tasks involved in both running the boat and operating strike planes can only be mastered through constant training and experience, which distinguishes the aircraft carrier from something like a missile. This is why countering the carrier, in the form of submarines and anti-ship missiles, has become a vastly more preferred strategy for most of the countries of the world than attempting to adopt carrier warfare or even acquiring fleet aircraft carriers.

Chapter 4

THE NUCLEAR REVOLUTION

AT AN ARMISTICE DAY ceremony on November 11, 1948, during an address as part of the events, General Omar Bradley stated that "ours is a world of nuclear giants and ethical infants." What did he mean? He was likely referring to the tremendous, and not completely understood, destructive capacity of the atomic bomb, and what it meant for the moral and ethical dilemmas common to warfare. For leaders who came of age in World War II, where authorizing an air strike meant sending hundreds of planes to drop tens of thousands of bombs, the idea that one plane carrying one bomb could deliver a destructive blow equivalent to literally thousands of conventional bombs was revolutionary.[1] Scientists, officers, and strategists—including prominent Manhattan Project leaders like Robert Oppenheimer, professors like Bernard Brodie, and generals like Bradley—believed that nuclear weapons represented a unique era of fighting that required reexamining the very way society conceptualized warfare.

More than sixty years later, nuclear weapons are still the coin of the realm in international power politics. In the United States, preventing the proliferation of weapons of mass destruction, but especially nuclear weapons, has been at the top of the foreign policy agenda for decades (Bush 2006).

Given that nuclear weapons have not been used in war since 1945, it is curious that successive U.S. administrations have placed such a high premium on preventing the proliferation of weapons of mass destruction in general and nuclear weapons in particular. The explanation likely lies in the belief that nuclear weapons as an element of national power give states important coercive capabilities. How is it that North Korea, with its tiny economy and backward regime, can make the front page of the *New York Times*? The knowledge that North Korea has nuclear weapons, and could sell them to terrorists or use them in a war, has elevated North Korea on the list of U.S. foreign policy priorities far above where it would otherwise exist. It is both the impact on the coercive power of states and the potential for actual use that makes nuclear weapons so crucial.

Evaluating nuclear weapons through the lens of adoption-capacity theory exposes a slightly different view of the factors that drive state decision making in the nuclear age than that of traditional theories. Most of the prominent literature on the spread of nuclear weapons along with more current discussions

---

[1] Several books have utilized the phrase "the nuclear revolution." See Benthem van den Bergh 1992; Kintner and Scott 1968; Mandelbaum 1981.

of the proliferation risks from countries like Iran revolves around how the international community can influence the desire of states to acquire nuclear weapons. In contrast, assessing the full range of strategic choices that states can make in response to nuclear weapons reveals the critical significance of financial and material constraints on their acquisition. Those countries that make the news for nuclear-related reasons tend to be those already pretty close to acquiring nuclear weapons. But to focus only on those countries in the news misses the much larger number of countries that do not acquire nuclear weapons—since news coverage is subject to a selection effect. While the desire to acquire nuclear weapons is clearly relevant, the actual ability to adopt given financial constraints plays a large role in determining the strategic choices states make in response to nuclear weapons.

This chapter makes three essential arguments about the spread and impact of nuclear weapons. First, the exceptional nature of the potential destruction from nuclear weapons means even rudimentary nuclear weapons are still instruments of enormous power. While there are organizational challenges associated with nuclear weapons, one factor that distinguishes them from other innovations is the incredible importance of the physical weapons system independent of any integration into a national military structure. Second, the enormous level of financial intensity necessary to acquire nuclear weapons has always functioned as a significant constraint on the diffusion of nuclear weapons. The inherent power of the atomic bomb, however, means that unlike most military systems, states can cumulatively invest over time in even a rudimentary nuclear weapons development program. This has critically aided states beyond the major powers in acquiring nuclear weapons. Third, the nuclear weapons case offers some clear examples, such as Japan, South Korea, Australia, and others, of states with the capability to build nuclear weapons that have chosen to bandwagon with the first mover, the United States, and pay alternative costs in the form of obligations within alliances, rather than pay the physical cost of weapons development.

Nuclear Weapons as a Military Innovation

Few military technologies have had as large an impact as nuclear weapons. Beginning with Brodie's *The Absolute Weapon* (1946), most scholars have maintained that nuclear weapons go far beyond conventional weaponry in the magnitude and speed with which escalation could occur, changing the character of warfare. Some have also claimed that nuclear weapons are defining weapons in international politics (Betts 1987; Jervis 1988, 1989; Mandelbaum 1981).[2]

[2] Bernard Brodie and Fawn McKay Brodie (1973, 233) called nuclear weapons "the most revolutionary military development of modern times." The belief that nuclear weapons have had a

Throughout the cold war, and focusing mostly on the United States and the Soviet Union, international relations scholars evaluated the relative importance of nuclear weapons in deterring the threat of conventional and nuclear war (Betts 1987; Jervis 1989; Schelling 1960; Snyder 1961). These studies included extensive debates about the utility of "rational deterrence theory," which views states as unitary actors rationally weighing costs and benefits.[3] A growing body of empirical statistical research also emerged on the influence of nuclear weapons (Bueno de Mesquita and Riker 1982; Jervis 1986; Vasquez 1991). Scholars also posited that nuclear weapons played an important role in proxy conflicts involving allies of the United States and the Soviet Union, prompting research on extended deterrence and its relative effectiveness (Huth 1988; Huth and Russett 1984). More recent research has continued to probe the importance of nuclear weapons for the success of extended deterrence promises (Danilovic 2002; Fearon 1994b; Signorino and Tarar 2006).

The existing academic literature on the spread of nuclear weapons focuses most on the factors that influence state interest in acquiring nuclear weapons, especially external security threats and domestic politics (Sagan and Waltz 1995). Sonali Singh and Christopher Way (2004) find that both security and economic factors influence the probability of proliferation. They use survival analysis and multinomial logit to estimate the probability that a state initiates a nuclear weapons program, pursues the acquisition of nuclear weapons, and actually acquires nuclear weapons.

Scott Sagan (1996) argues that the drive to acquire nuclear weapons is heavily influenced by competing domestic interest groups, rather than international security concerns. Nuclear weapons development is influenced by bureaucratic political debates, including civil-military struggles for control as well as a desire by interest groups within countries to support perceptually dominant international norms, including nonproliferation norms (85–86). James Walsh (1997) similarly emphasizes the importance of bureaucratic politics, while Jacques Hymans (2006) contends that national identity concerns are a critical factor in predicting nuclear weapons decision making.

Are Nuclear Weapons a Military Innovation?

Nuclear weapons are one of the largest, if not the largest, innovations in the production of military power in recorded history.[4] The splitting of uranium or

---

revolutionary impact on international security is not universal (Mueller 1999a, 272–73; Mueller 1999b).

[3] One influential debate about rational deterrence theory appeared in a special issue of *World Politics* in 1989 (vol. 42, no. 1).

[4] Some of the best work on whether nuclear weapons are a military innovation is by Evangelista (1988).

plutonium isotopes for atomic bombs represented a clear step-change increase in the possible destructive power accomplishable with a single bomb.[5] The additional health hazards caused by the massive release of radioactive energy at the blast site only added to the destruction caused by nuclear weapons.

This is not to say that all nuclear weapons are more powerful, based on explosive tonnage, than all conventional weapons or especially the combination of multiple conventional weapons. The firebombings of Dresden and Tokyo in World War II by Allied air forces generated levels of destruction comparable in some ways to Hiroshima (twelve to fifteen kilotons) and Nagasaki (twenty to twenty-two kilotons). Yet the firebombing of Dresden required thousands of bombers dropping thousands of bombs, while the Hiroshima bombing required one plane dropping one bomb. The largest conventional bombs, like the BLU-82 "Daisy Cutter" bomb, do have yields larger than the smallest-known nuclear warheads, like the M-388 "Davy Crockett," which had a yield of .01 kilotons. That a small nuclear device had a yield comparable to one of the largest conventional devices shows the staggering increase in explosive power offered by nuclear weapons.

The use of nuclear weapons shifts the organization of military power because it changes the range of possible ways a state can destroy a target. For example, if the same destructive force that required a thousand bombers now requires one bomber, states may change the way they think about designing bombers, shifting from quantity to quality in production. Alternative delivery options, like missiles, can also influence force planning.

Nuclear weapons fall outside the normal parameters of debate about how to integrate new weapons because they do not necessarily fit into any particular uniformed service structure, unlike, say, battleships and the navy. Harvey Sapolsky, for instance, highlights how the uniformed services in the United States each attempted to "capture" nuclear weapons due to their cold war budgetary importance. The resulting interservice rivalry led each service to create options for nuclear-tipped missiles, resulting in the land-launched intercontinental ballistic missile and the ballistic missile submarine (Sapolsky 1972, 1996). But this is not the only solution.

While the different services can and do jockey for control of nuclear weapons, it is also possible to operate a nuclear arsenal outside the normal command structure, as the Soviet Union showed. Additionally, since no one really knows what a nuclear war would be like, it is difficult—and this was more true in the early nuclear age—for military officers to claim special expertise or knowledge over civilians, especially since civilian scientists were the first to develop the weapons (Rosen 1991, 21–22). Regardless of where nuclear weapons

---

[5] This was followed by the development of thermonuclear bombs, or combinations of hydrogen or hydrogen-related isotopes undergoing fusion reactions triggered by fission reactions (Rhodes 1995).

are located within a national bureaucratic structure, their existence sharpens budget debates because nuclear weapons are also extremely expensive compared with other weapons.

One could say that nuclear weapons do not count as a military innovation because their importance lies almost entirely in the technology of the nuclear explosion, meaning the organizational components are not just small but instead insignificant. There are a few reasons that this argument is not persuasive. First, the vast differences in the way states have deployed nuclear weapons—from the United States' nuclear "triad" to China's minimal nuclear arsenal to India's separation of nuclear warhead components from each other and any possible delivery mechanisms—show that organizations do matter in managing nuclear weapons. Second, even if the importance of nuclear weapons lies mostly in their technological aspects, the degree of the relative difference between nuclear and conventional weapons is so large that the technology itself represents a shift in the production of military power. No other case qualifies for this designation, making nuclear weapons a special case of a military innovation.

The Debut of Nuclear Weapons

The beginning of the nuclear age from a diffusion perspective is August 1945. On August 6, 1945, a U.S. B-29 bomber, the *Enola Gay*, dropped the "Little Boy," an enriched uranium gun-type device, on the Japanese city of Hiroshima. Three days later, the "Fat Man" plutonium implosion device was dropped on the Japanese city of Nagasaki. These two bombs are still the only nuclear weapons ever used in warfare. The atomic bomb was immediately heralded as a war-changing weapon, an exceptional form of military power (Brodie 1946; Groves 1962).

Nuclear weapons break the mold in thinking about the difference between the incubation period for an innovation and the demonstration point at which diffusion begins. The nuclear incubation period was almost nil given the vast increase in possible destructive power through the mere introduction of the technology into armed combat. The technology itself was sufficiently different from previous military technologies as to make it a mature innovation on debut, even without fully formed operational concepts.

The simultaneous emergence of nuclear weapons technology and nuclear weapons as an innovation, in contrast to cases like carrier warfare, requires compressing some of the theoretical elements described in the other chapters. For example, the question of information concerning the importance of the innovation is not as applicable. While there are some statements by leaders like Mao Tse-tung about the irrelevance of nuclear weapons for international security, nearly all global leaders instantly grasped the significance of them. If Mao's opinion of nuclear weapons is judged by his actions rather than his words, his diversion of scarce funds into the development of a Chinese nuclear weapon shows their relevance for international politics (Johnston 1995; Lewis 2007). But this does not mean that transmission of all information concerning the innovation was simultaneous. There is and was a big difference between knowing that nuclear weapons are important and knowing how to build a nuclear bomb.[6]

Predicting the Spread and Impact of Nuclear Weapons

This section applies adoption-capacity theory to the nuclear weapons case in order to make predictions about the diffusion of nuclear weapons, state responses, and the impact of the nuclear weapons innovation on international politics.

Required Financial Intensity

By all measures, acquiring nuclear weapons is an intensive financial process. Even excluding the cost of potential delivery vehicles like ICBMs, bombers, and/or submarines, the cost per unit of a nuclear weapon is extraordinarily high. For example, the Manhattan Project in the United States cost $20 billion in constant 1996 dollars.[7] During the Manhattan Project, U.S. scientists also had a great deal of uncertainty about how to design a nuclear bomb, which led to a higher degree of experimentation and drove up the costs. As information on atomic designs has spread, especially early atomic designs, the relative cost of building a nuclear bomb has declined. But the cost is still enormous. Stephen Schwartz finds that the United States, through 1996, spent about $400 billion in constant 1996 dollars on developing nuclear weapons. Comparing the cost of the Manhattan Project to the B-29 Superfortress, a strategic bomber designed to deliver conventional bombs that also served as the first U.S. nuclear delivery vehicle, demonstrates the relative unit costs of nuclear weapons. The U.S. government purchased a total of 3,960 B-29s at a cost of about $639,000 per plane. The total cost of over $2 billion for every B-29 purchased by the U.S. government is only about 13 percent the cost of the Manhattan Project alone (Pike 2005b; Schwartz 1998).[8]

---

[6] There are also material constraints, as discussed below.

[7] If the cost of delivery vehicles for nuclear weapons is included, the unit cost of a nuclear arsenal is even higher. Stephen Schwartz (1998) finds that delivery vehicles and associated costs comprise 86 percent of total spending on the U.S. nuclear arsenal from 1940 to 1996, or about 4.7 trillion dollars. These costs are not inevitable, but states must choose to pay for either reliability or the costs of an alternative delivery mechanism like smuggling.

[8] Additionally, states interested in building nuclear weapons have to acquire enriched uranium, acquire regular uranium and design an enrichment facility, and/or acquire plutonium. Global supplies of nuclear materials are limited and regulated by the existing nuclear powers, creating further costs for potential proliferators.

There are also significant technical barriers to adoption. Countries need to develop or acquire enriched uranium or plutonium. They also need to create a significant technical infrastructure. Recent research by Singh and Way (2004) as well as Matthew Fuhrmann (2009) shows that these technical barriers are a big obstacle to acquiring nuclear weapons.

Still, the nuclear weapons case also highlights how the required financial intensity for adoption can shift over time, demonstrating some dynamism in the concept. The cost per unit of nuclear weapons is likely lower, in a relative sense, for states today than during the Manhattan Project for two reasons. First, as the technological first mover, the United States invested in many dead-ends and could not master the most efficient method of enriching uranium: centrifuge enrichment. Especially after the release of technical details concerning the Manhattan Project—both directly transferred from the United States to allies and through public information—future adopters could then take advantage of this experience, helping them to avoid pitfalls, and zero in on the best and cheapest method of acquiring nuclear weapons.

Second, the cumulative nature of investments in nuclear weapons creates a limiting factor on the financial intensity level required for adoption. With some innovations like battlefleet warfare, adoption requires the integration of new organizational methods and the purchase of technologies with a limited shelf life. Recall the concern of the British Admiralty in the years before World War I about the shrinking life cycle of its battleships and the greater investments required over time for the purchase of each generation of ships. Investments in the technological aspects of the innovation were not cumulative; they were sunk into each ship, which then went out of service in a number of years and was replaced by another ship, which in turn required investing in even newer technologies.

Nuclear weapons are different because an arsenal of just one rudimentary nuclear weapon can influence international politics.[9] Unlike other types of weapons, states can invest in the components of nuclear weapons over time with confidence that the culmination of their investment, a nuclear weapon, will not become obsolete. First-generation nuclear weapons built after a few decades of incremental investment in 2010 provide a lot more relative power than a "new" battleship produced in 1910 based on designs from the 1860s or even the 1890s. Since even rudimentary nuclear weapons maintain more value than similarly aged conventional weapons, the investment pattern for those weapons is different than for weapons whose value declines more steeply with age.[10]

---

[9] While great powers might face competitive pressures that force them to build up their arsenal, this refers to the influence from first demonstrating a nuclear capability.

[10] There are reliability questions since basic nuclear weapons, especially if untested, may possess less deterrent value than weapons whose reliability is more assured. Unlike with artillery, though, where reliability questions may make them useless from a battlefield perspective, even an uncertain nuclear weapons capability can have enormous coercive implications.

While there are now civil applications for nuclear power, at the time of its creation, nuclear power had an entirely military purpose, making it more akin to the rifle than the railroad. The railroad had commercial applications prior to military ones, while the rifle existed for solely military uses. Recall that whether the technology underlying innovations is mostly military or mostly civilian is a function of the technology at the debut point of the innovation. It is true that the dawning of the nuclear age also led to the creation of a commercial nuclear power industry. But unlike the railroad, there are no strictly commercial prospects for nuclear *weapons*, even if there are commercial applications for nuclear power, a related technology. So nuclear weapons have both an enormous unit cost and a core military purpose, meaning obtaining a nuclear weapon should require an extremely high level of financial intensity, although there are some mitigating factors.

Required Organizational Capital

In contrast to the high financial intensity level associated with nuclear weapons, the organizational requirements for adopting nuclear weapons are much more limited. Given the slow spread of nuclear weapons over time, it might seem puzzling to call the organizational constraints limited. Yet the unit-level destructive envelope of nuclear weapons is well beyond the experience of any military organization (Rosen 1991). Therefore, examining the level of organizational capital required to adopt the nuclear innovation requires a small deviation of sorts from the framework used in the other empirical chapters.

Innovations that require high levels of organizational capital to implement generally involve mastering detailed tasks that are harder for some organizations than others. In contrast, while nuclear weapons are complex, there are multiple paths to their development, from focusing on plutonium and/or uranium to different methods of uranium enrichment. There are also many different ways to bring nuclear weapons into national military plans because nuclear weapons, in theory, do not necessarily bolster the bureaucratic position of any particular military service or way of thinking about force employment. It is possible that acquiring nuclear weapons will lead to the privileging of one military institution over another if a state chooses a single means of delivering nuclear weapons. Alternatively, a state could spread responsibilities for nuclear delivery vehicles across military services. Another option is for the state to take decisions concerning nuclear weapons out of the hands of the traditional military and develop an alternative command structure to deal with nuclear weapons. It is difficult to predict how the organizational issues associated with nuclear weapons will resolve themselves in a military organizational context, because nuclear weapons are beyond the specialties of organizational interest groups like fighter pilots or tank drivers, rather than obviously favoring one group over another.

The distinctiveness of the nuclear weapons development process also lowers the importance of experimentation precursors for innovation purposes, as we should expect for those innovations requiring lower levels of organizational capital to adopt. The basic science underlying the development of nuclear weapons is so sufficiently distinct from other weapons that experimentation in regular military areas is likely unrelated to nuclear weapons development.

Since the possession of nuclear weapons is enough to shift the way a state is considered in international politics, a state with extremely low levels of organizational capital can extract nearly the same benefits from the technology that a state with a higher level of organizational capital would gain.[11] Essentially, a state with nuclear weapons and a totally dysfunctional military organization is still a nuclear weapons state, and it will be treated as such in international politics regardless of whether its military is adept at antisubmarine warfare or tank maneuvering. Contrast this with the tank. Different armies have employed tanks in different ways, and those differences, as Germany poignantly showed in the opening campaigns of World War II, can play a major role in determining the success or failure of an army, even when it confronts another army with tanks.

This conception of the impact of nuclear weapons is somewhat different from that proposed by Keir Lieber and Daryl Press (2006a, 2006b), who focus on the ability of the United States to deliver a first strike against Russia or any other nuclear state. But the evidence argues for a broader approach. For one, most of the states that attempt to acquire nuclear weapons do not do so with the goal of developing an arsenal to compete with the United States. States leverage nuclear weapons for regional security reasons, to deter an attack from the United States or a larger nuclear power, or for other reasons. Outside the U.S.-Soviet/Russian context, nuclear weapons possession still matters as a symbol of power and a coercive bargaining chip. Otherwise, U.S. policymakers would not worry about nuclear proliferation because the small arsenals of new nuclear powers would not threaten the United States. Yet possessing even a single nuclear weapon influences America's strategic calculations and seems to make coercive success harder.[12] Second, since from a diffusion perspective what matters is the view of potential adopters, the fact that other countries highly value small numbers of nuclear weapons proves the importance of the

---

[11] A more organizationally capable state might be better at designing procedures to effectively leverage its nuclear weapons for influence and/or build more nuclear weapons. The difference between having and not having nuclear weapons is still much larger than the relative gap between nuclearized states. There is a reason that people call it the "nuclear club."

[12] For example, some argue one reason that the United States was able to invade Iraq but not North Korea was confidence that Iraq did not have a nuclear arsenal, in contrast to concern about North Korea's nuclear arsenal, which lowered the expected costs of invading Iraq and raised those of invading North Korea.

weapons. In the scope of nuclear history, the competition between the United States and the Soviet Union increasingly appears like the exception rather than the rule (Rosen 2006).

The first-mover test, evaluating the organizational capital level of the United States during World War II, presents some conflicting results. While the United States had a higher organizational age than newer states like the Soviet Union or recent war losers like Germany, it had a lower organizational age than Great Britain. Additionally, the U.S. military lacked a general critical task in the pre–World War II period, though the carrier chapter demonstrates that the United States had a broadly conceptualized critical task in the area of naval warfare. Finally, the United States invested heavily in experimentation in the form of the Manhattan Project, but had not heavily experimented beyond naval warfare in the prior decade, likely at least in part due to the Great Depression. So there are confounding factors that make the signal from the first-mover test less clear than in some cases. The first-mover test is especially unclear due to the uniqueness of nuclear weapons. Since the atomic bomb was so far outside the envelope of normal military developments, it is possible that the first-mover test would be less accurate than in some of the other cases.

Given great differences in the potential pathways to proliferation—there is less need to specifically modify a military organization on to a particular pathway to take advantage of nuclear weapons—the level of organizational capital required for adoption is low. Nation-states with highly dysfunctional military organizations possessing low levels of organizational capital should still be able to acquire nuclear weapons.

Nuclear Weapons Diffusion Predictions

Based on the general requirement of high levels of financial intensity but low levels of organizational capital for adoption, the basic diffusion prediction is that, ceteris paribus, wealthy, powerful states should acquire the innovation over time, though slowly given the high costs, and the innovation should continue spreading slowly over time, especially as the diffusion of tacit knowledge lowers the bar for new adopters. As described above, one distinction between the nuclear weapons case and others requiring high levels of financial intensity for implementation is the cumulative nature of investments in nuclear weapons. This should make the acquisition of nuclear weapons more plausible for a broader range of states, albeit over a long time frame. Since the innovation requires low levels of organizational capital to adopt, the innovation is sustaining. This means adoption should generally become more likely as organizational age increases or no relationship should exist.

Other Strategic Choices in Response to Nuclear Weapons

In the nuclear weapons case, states responding to the first mover should consider the full range of response options, from shifting toward neutrality to countering the innovation through the development of alternate military means to balancing or bandwagoning to adoption of the innovation. The intense financial attributes required for adoption should make alliance strategies probable. States may want to form an alliance with a nuclear power in the hopes of acquiring nuclear technology over time or guaranteeing protection from a nuclear power in case they are threatened. The especially high unit costs of developing and possessing nuclear weapons mean that states that might ordinarily be late adopters of military innovations might feel pressure to make faster strategic decisions in favor of alliances in addition to any procurement path, because they cannot afford the security risk from a lack of nuclear "coverage." This means that, depending on the preexisting relationships a given state has with the first mover and other potential adopters, many states might be tempted to engage in dominant alliance strategies.

As the civil defense measures taken by the United States, the Soviet Union, and other states—in the 1950s and 1960s especially—show, there are a variety of actions that states can take when faced with a nuclear threat besides acquiring nuclear weapons or seeking an external alliance. Spreading industrial sites around a country as well as hardening key facilities and bunkers beyond the penetration capacity of nuclear weapons are two ways a state can try to mitigate the societal impact of a nuclear attack. To prevent a nuclear attack from reaching its soil, states might also develop defenses to shoot down ballistic missiles and/or nuclear-armed bombers. Nevertheless, the explosive power and radiation effect of nuclear weapons will make full mitigation difficult, so countering is unlikely to be a dominant response strategy. Nearly all states, whether they acquire nuclear weapons or not, should engage in some forms of countering to try to minimize the impact of nuclear weapons used against them.

The nuclear weapons case also includes a concerted effort by the international community to employ treaty mechanisms, mainly the Nuclear Non-Proliferation Treaty (NPT), to constrain nuclear proliferation. The existence of an international treaty regulating the proliferation of the weapons system makes the nuclear weapons diffusion case distinct from previous ones in which such treaties did not exist.[13] The NPT explicitly outlaws the acquisition of nuclear weapons by all signatory states that do not already have nuclear weapons, but legalizes the nuclear arsenals of those states that had nuclear weapons in July 1968, when the treaty opened for signature. It does not limit their acquisition of more nuclear weapons.

---

[13] Kellogg-Briand and other arms control measures of the 1920s and 1930s regulated arsenal stockpiles, not the acquisition of the weapons in the first place, like the NPT.

NPT proponents like George Bunn (1994), Leonard Spector (1992), and others argue that the NPT has an independent causal impact on the probability of nuclear proliferation by generating an international norm against acquiring nuclear weapons, and guaranteeing diplomatic costs for those that violate the norm. Proponents also claim that the NPT transforms national interests by generating powerful interest group coalitions in favor of compliance, making it less likely that a decision to acquire nuclear weapons will appear favorable in domestic political debates (Chayes and Chayes 1995; Sagan 1996).[14]

In contrast, scholars like Colin Gray (1992, 222–25) make arguments in line with John Mearsheimer's critiques (1994) of international institutions in general, asserting that the NPT adherence is less a reflection of the independent causal influence of arms control agreements than a statement that some nation-states do not want to acquire nuclear weapons. If states have incentives to develop a prohibited capability for national interest reasons, the costs of compliance increase, and treaty adherence will decline (Downs, Rocke, and Barsoom 1996).

States likely use the NPT as a strategic supplement rather than a dominant strategy. Given a world of imperfect treaty adherence, it seems unlikely that a state would reorient its national strategy around trust that the NPT will completely prevent proliferation. The NPT simply helps states limit the downside risks to neutrality by controlling how many other states will have nuclear capabilities.[15]

Nuclear Weapons Impact Predictions

Adoption-capacity theory predicts that the high financial but low organizational costs of adoption mean that nuclear weapons should reinforce preexisting power balances in the period immediately following the debut of the innovation. In general, major powers should benefit from nuclear weapons. The innovation will not generate significant opportunities for new states to gain in relative power in comparison with major states, though minor powers that do acquire nuclear weapons will experience large relative power gains.

---

[14] It is also possible that the NPT has shifted global preferences and generated a dominant norm against the possession of nuclear weapons that serves as an additional barrier to adoption (Tannenwald and Price 1996). Normative opprobrium could function in the nuclear case to deter potential proliferators because of the "cost" of violating the norm. Whether the NPT is effective for strategic reasons, normative reasons, or a combination of the two, however, the causal arrows point in identical directions, making the arguments difficult to empirically distinguish.

[15] Then again, if the neutral state is geographically close to an existing nuclear state, an effective NPT might constrain its options by preventing other states in the region from developing nuclear weapons, and allowing the neutral state to shift to a bandwagoning or balancing strategy. These possible reactions, which go in opposite directions but are both responses to the NPT, make it difficult to develop predictions about the independent impact of the NPT on strategic choices.

The uniqueness of nuclear weapons should also lead to large first-mover and early adopter advantages. While the innovation should spread over time, the relative advantages provided by nuclear weapons should supply a big relative power boost, since the spread of nuclear weapons should be fairly slow, even to major powers, because of the exceptionally large financial costs and material constraints for early adopters.[16] As discussed above, though, since it is possible to invest over time and build a nuclear program, the cost per unit of nuclear weapons should decline over time, making it relatively easier for late adopters to acquire nuclear weapons.

Adoption-capacity theory also makes a number of predictions with regard to the relationship between innovation diffusion, power, and war that should apply in the nuclear case. Nuclear weapons should allow for power projection at greater distances by lowering the costs of coercion once the weapon is acquired, potentially shifting the geographic range of the possible for states that acquire nuclear weapons. This could mean that states will become involved in more militarized disputes, since their capabilities now extend further than before.

Nuclear Diffusion Research Design

Testing the hypotheses developed above requires evaluating two different sets of predictions: those regarding the spread of nuclear weapons, and those regarding their effect on the international security environment. The acquisition of nuclear weapons is tested using a data set organized in country-year form, meaning that there is an observation for every country in every year of its existence. The country-year setup yields a total of 7,925 observations from the nuclear age, from 1946, the first year after the United States acquired nuclear weapons, through 2002, the last year for which data on most of the variables are available.[17]

Nuclear Diffusion Dependent Variable

The dependent variables for the nuclear proliferation tests involve a two-stage measurement scheme separating whether a country has a nuclear weapons program from whether a country actually acquires nuclear weapons. Given the number of states with nuclear weapons programs that have not acquired nuclear weapons, including Argentina, Brazil, and Sweden, it makes sense to think

---

[16] Even if late adopters have lower cost barriers to entry, they also face operational barriers and smaller arsenals initially. So how nuclear weapons actually function in international politics, especially whether arsenal size matters, changes the extent to which a late adopter advantage might exist.

[17] Other recent studies of nuclear proliferation, by Singh and Way (2004) as well as Dong-Joon Jo and Erik Gartzke (2007), have also utilized a similar cross-sectional time-series country setup.

about the capabilities of countries as more complicated than simply whether or not a state has nuclear weapons.

The nuclear program variable, *Nuclear Program*, measures whether or not a state in any given year has an active nuclear weapons development program. The variable is coded a 1 if a state has an active nuclear weapons development program and a 0 otherwise. An active program is defined based on the decision by a country to invest funds in a nuclear weapons development program or otherwise authorize steps toward acquiring nuclear weapons, like setting up a part of the military designed to handle nuclear weapons, or developing a plan to procure and/or develop enriched uranium or plutonium (Singh and Way 2004, 867).

The second nuclear acquisition variable, *Nuclear Weapons*, measures whether or not a country in a given year has nuclear weapons. The country is coded a 0 if it does not have nuclear weapons and 1 if it does. While the possession of nuclear weapons is often considered a relatively simple question—either a state has a nuclear weapon or does not—there are some differences in the way that scholars have coded the nuclear arsenals of states over the last fifty years. For example, while some have coded Israel as acquiring nuclear weapons in 1966, others say it was 1972 or even 1973. While some argue that India acquired nuclear weapons capabilities in 1974 when it exploded a peaceful nuclear device, others maintain it was not until 1988, 1990, or even 1998 that it finally acquired nuclear weapons.[18] Table 4.1 shows the coding of nuclear weapons possession and includes the citations used to determine the coding for each case.

Adoption-Capacity Theory Variables

The independent, or explanatory, variables described below are the quantitative conversions of adoption-capacity theory for the purpose of statistical testing. If financial intensity is the crucial capabilities-based constraint that governs the probability of proliferation, because the organizational capital required to implement the innovation is low, then the countries most likely to adopt nuclear weapons are those that have a high level of financial capacity. The first two variables are drawn from GDP data. GDP per capita is a strong measure of overall economic development and specifically the sophistication of economic development. To account for the possibility that the necessary level of sophistication is a threshold effect, rather than a linear relationship, I followed Singh and Way by including a squared measure of GDP.[19]

---

[18] In general, the acquisition of nuclear weapons capabilities is defined as the explosion of a nuclear device. For a few cases like Israel and South Africa, the definition is less appropriate. Following Singh and Way (2004), the post–Soviet states are excluded because they did not actively attempt to acquire nuclear weapons. Including them does not substantively change the results.

[19] GDP data are from Penn World Tables, Angus Maddison, and Kristian Gleditsch (Singh and Way 2004, 867–68).

TABLE 4.1
Nuclear Weapons States, 1945–2002

Country	Years possessing nuclear weapons	Sources
United States	1945–present	Natural Resources Defense Council 2002
Soviet Union/Russia	1949–present	Natural Resources Defense Council 2002
United Kingdom	1952–present	Natural Resources Defense Council 2002
France	1960–present	Norris 1994
China	1964–present	Natural Resources Defense Council 2002
Israel	1967–present	Cohen 1998
India	1988–present	Natural Resources Defense Council 2002; Perkovich 1999
South Africa	1979–91	Albright 1994
Pakistan	1987–present	Nizamani 2000; Perkovich 1999
Not included in data set		
Kazakhstan	1991–95	Bennett and Stam 2004
Ukraine	1991–96	Bennett and Stam 2004
Belarus	1991–96	Bennett and Stam 2004

Note: North Korea is coded as acquiring nuclear weapons in 2003, falling outside the period of study for data-availability reasons.

Yet GDP information may not be enough to account for financial intensity since it does not explicitly account for industrial capacity. I therefore included three more measures. The first is a reindexed measure of the energy, iron, and steel production data from the COW Projection Material Capabilities. To account for measurement problems in the data, I adopt a similar solution to that employed by Dong-Joon Jo and Erik Gartzke by indexing the energy, iron, and steel production capacity of a country in a given year as a percentage of how it relates to the energy, iron, and steel production of the international system in a given year.[20] The next capacity measure is that of the nuclear capabilities data utilized by Jo and Gartzke (2007, 172–73) based on seven industrial characteristics of each country: "uranium deposits, metallurgists, chemical engineers, and nuclear engineers/physicists/chemists, electronic/explosive specialists, nitric acid production capacity, and electricity production capacity." Unfortunately

[20] The equation is ((Energy/(Sum)Energy)+(Iron and Steel/(Sum)Iron and Steel))/2. The results are significant with multiple specifications, including an unindexed variable.

this measure only runs through 1992, so I present all results both with and without this variable.

Finally, it is possible that there is a relationship between military spending and adoption, since those countries that spend more on their militaries are more likely to have both the financial capacity and military interest to develop nuclear weapons. COW military expenditure data from 1914 onward are based on current year U.S. dollars (Correlates of War 2 Project 2006, 20).[21] I use military spending data drawn from COW as a primary measure of military investments. I also created an indexed measure of military spending based on the $20 billion cost for the Manhattan Project in 1996 U.S. dollars, turning that figure into current year U.S. dollars for every year between 1945 and 2002. That cost is divided by the military spending for a country in a given year, since it is also measured in current year U.S. dollars. Using the current year cost of the Manhattan Project divided by the defense expenditures of a country may better take inflation into account than military spending alone. The results are presented with the military spending variable, since it is drawn straight from the COW data, and is more comparable with past and future research. Including the indexed measure instead did not change the results.[22]

In combination with the other variables described above, these metrics offer a reasonable test of the financial intensity hypothesis. As described above, finding a significant relationship between these financial metrics and nuclear weapons proliferation would provide evidence in favor of adoption-capacity theory. Yet these measures would also offer some support for other theories, since realist and other theories recognize the importance of material capabilities. In this case, the point is not to reinvent the wheel by showing that adoption-capacity theory can provide an explanation for nuclear proliferation that no other theory can approach but to demonstrate that adoption-capacity theory can also explain the spread of nuclear weapons.

Two other variables also help measure the relationship between adoption capacity and nuclear weapons. The organizational capital level of states in the system is operationalized by measuring the "organizational age" of each state.[23] Organizational age is coded by measuring the time gap in each country year

[21] As with the rest of the COW capabilities data used in this book, they are drawn from version 3.02 (Singer 1987; Singer, Bremer, and Stuckey 1972).

[22] The numerator in the indexed variable changes from year to year based on inflation in current year U.S. dollars. Therefore, the percentage of defense expenditures required for the same country to implement the Manhattan Project could change from year to year depending on macroeconomic shifts. While not on its own a perfect indicator of financial intensity, it does provide additional leverage.

[23] Organizational age offers the best theoretically derived measure of organizational capital for this case since full data on experimentation and critical task focus are not available for some states in most years. This also creates comparability with the statistical tests in chapter 6, which use organizational age as a measure of organizational capital as well.

between the current data year and the last time the country lost a war, as defined by COW, or experienced a change in regime, defined as a change of three or more on the Polity scale (Marshall and Jaggers 2002; Sarkees 2000).[24] Regime changes, like from autocracy to democracy, along with crucial events like war losses, have the potential to lead to real organizational changes that eliminate hierarchies and "reset" the organizational age of a given military. The more recent the regime change or war loss is, the lower the organizational age.

The theory also predicts that there should be changes over time in the ability of states to acquire nuclear weapons due to the diffusion of information about nuclear weapons and the existence of growing numbers of scientists with nuclear experience, which lowers the financial intensity level required for adoption. A timing variable is created to test the role of both information diffusion and lowering costs through the log transformation of the number of years since 1945, when the United States tested a nuclear weapon and then used the atomic bomb in war (Jo and Gartzke 2007, 173).

Accounting for Existing Quantitative Literature on Nuclear Weapons

It is important to take seriously Christopher Achen's injunction (2002, 2005) that the inclusion of excessive control variables unrelated to theory can distort results. Therefore, all results are presented both with and without control variables to ensure the adoption-capacity variables are not artificially made significant through the inclusion of controls. There are several security-related factors existing research predicts should be related to nuclear proliferation. Controlling for them is crucial since measures like military spending might be significant not for reasons related to financial intensity but because security concerns cause a nation to boost its defense spending. The theory predicts that countries that ally with nuclear powers should be much less likely than other types of states to adopt nuclear weapons. Data gathered from the COW Formal Alliance Data Set (3.03) is used to code whether or not a state has an alliance in a given year with a nuclear power (Gibler and Sarkees 2004). Given that only actual commitments to come to the aid of a state in the event of an attack should influence the way a state evaluates the utility of acquiring nuclear weapons, only defense pacts are included.[25] The *Alliances* variable is coded a 1 if a state has an defense pact with a nuclear power in a given country year and a 0 in other cases.

---

[24] Remeasuring the variable using the major war scale described in chapter 2 does not influence the results.

[25] This limits the set of alliances to those with explicit security guarantees. The COW alliance data are used to mimic the data used in previous research and increase the comparability of this study. This excludes neutrality/nonaggression and entente pacts (Jo and Gartzke 2007, 174; Singh and Way 2004, 860).

Most theories of proliferation argue that at least some level of external threat is necessary to generate domestic support within a country for acquiring nuclear weapons (Mearsheimer 1984; Singh and Way 2004, 882–83). An additional dichotomous variable, *Nuclear Threat*, codes whether or not the enduring rivalry featured a nuclear-armed rival, since nuclear rivals could create insecurity and encourage a state to acquire nuclear weapons, or alternatively, create fear that leads states to back away from the nuclear option (Bueno de Mesquita and Riker 1982; Jo and Gartzke 2007, 174; Mandelbaum 1995; Waltz 1995). Enduring rivalries are one way to measure whether or not a state faces a persistent security challenge that might prompt it to consider the nuclear option. A second measure of security threats that also takes into account the possibility of enduring rivalry is a five-year moving average of the number of militarized disputes in which a country has been involved, named *Dispute Propensity*. Finally, based on prior research on enduring rivalries and the list generated by D. Scott Bennett, a *Security Threat* variable is coded a 1 if a state in a given country year is part of a rivalry and a 0 otherwise (Bennett and Stam 1996; Diehl and Goertz 2000; Singh and Way 2004).

Two other variables, *System Satisfaction* and *Regional Satisfaction*, generated using the S coding scheme designed by Curtis Signorino and Jeffrey Ritter (1999), measure the satisfaction of a state with the international system leader and the regional system leader, respectively. The variables, as modified by Bennett and Stam (2004), run from 0 to 1 (as opposed to -1 to 1), with 0 indicating dissatisfaction and 1 showing satisfaction.[26]

Finally, a *Democracy Level* control measures domestic political regime type and whether it influences the probability that a state acquires nuclear weapons. The Polity IV data set, which measures the democracy and autocracy levels of all states in the international system from 1800 to 2003, was used to construct a dichotomous measure of regime type (Jaggers and Gurr 1995). Regimes with a combined polity score of eight or more, meaning their democracy score (-10 to 10) minus their autocracy score (-10 to 10) was equal to or greater than 8, were coded 1, with all other states defined as nondemocracies and coded a 0.[27]

Nuclear Diffusion Results

Simple statistical analysis, displayed in figure 4.1, shows the overall pattern of nuclear weapons diffusion in the international system, differentiating between the two stages of nuclear weapons development: program and acquisition.

---

[26] These results are consistent using Bueno de Mesquita Tau-B scheme through EUGene as well.

[27] Variations in the particular democracy variable, making it continuous, or adding a joint democracy variable did not significantly influence the results.

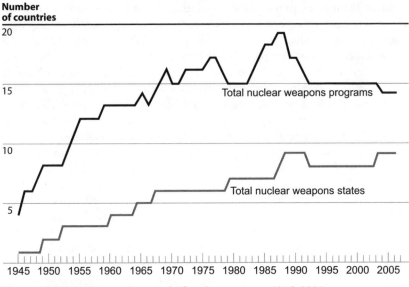

Figure 4.1. Initial data on the spread of nuclear weapons, 1945–2006.

Figure 4.1 depicts the general diffusion trend, which is similar to an external diffusion curve and consistent with the theoretical predictions. The number of adopters gradually increases over time, with a set of core adopting states in the early to midstages of diffusion, but then the spread of the innovation levels off.[28] The spread to nonmajor powers over time is expected, especially given the unique investment pattern possible for nuclear weapons, where states can invest smaller amounts on a yearly basis over time. Even a wealthy state like Israel, however, took about two decades to develop nuclear weapons, illustrating the large financial and other barriers to adoption. Many of the states that initiated programs to acquire nuclear weapons over the last fifty years did not achieve the programmatic goal of possessing nuclear weapons.[29] The gap between total programs and states with nuclear weapons validates thinking about the diffusion of nuclear weapons as a two-stage process. The divergent patterns indicate that different functions may govern the decision to initiate a nuclear weapons program and acquire nuclear weapons, meaning an independent statistical treatment of each stage is necessary. Table 4.2 presents three simple probit models

[28] If the diffusion of nuclear weapons was dominated more by inside information procured through official channels, a more S shaped curve would be likely.

[29] Some states abandoned these programs by choice, and others did so due to a lack of resources.

TABLE 4.2A
Predicting the Diffusion of Nuclear Weapons in Stages: Creating a Nuclear
Weapons Development Program

	No latent nuclear variable, no controls	Latent nuclear variable, no controls	No latent nuclear variable, controls	Latent nuclear variable, controls
	Model 1a	Model 1b	Model 1c	Model 1d
GDP per capita	0.0001	0.0001	0.0002	0.0001
	(0.0001)***	(0.0001)	(0.0001)***	(0.0001)*
GDP squared	$-1.49E-08$	$-9.29E-09$	$-1.37E-08$	$-1.25E-08$
	$(3.98E-09)$***	$(4.68E-09)$**	$(3.36E-09)$***	$(3.86E-09)$***
Industrial capacity	11.501	0.040	-2.310	-4.718
	(9.278)	(6.779)	(8.807)	(5.358)
Latent nuclear capacity		0.298		0.242
		(0.085)***		(0.091)***
Military spending	$1.08E-07$	$9.80E-08$	$5.89E-08$	$6.28E-08$
	$(2.53E-08)$***	$(2.35E-08)$***	$(2.28E-08)$**	$(2.55E-08)$**
Diffusion	-0.361	-0.432	-0.325	-0.310
	(0.182)**	(0.248)*	(0.293)	(0.346)
Organizational age	0.015	0.018	0.017	0.019
	(0.005)***	(0.006)***	(0.005)***	(0.006)***
Military personnel			0.002	0.001
			(0.001)**	(0.001)
Polity			0.014	0.011
			(0.017)	(0.017)
Nuclear ally			-0.336	-0.512
			(0.313)	(0.344)
Nuclear threat			0.055	0.095
			(0.524)	(0.495)
Dispute propensity			0.087	0.114
			(0.126)	(0.116)
Security threat			0.5	0.363
			(0.316)	(0.357)
Global satisfaction			0.811	0.585
			(0.859)	(0.946)
Regional satisfaction			-0.373	-0.168
			(0.269)	(0.322)
Constant	-0.856	-1.46	-1.933	-2.532
	(0.607)	(0.696)**	(0.967)**	(0.974)***

Notes: Robust standard errors in parentheses. Model 1a: N: 5,971, Wald chi2(6): 44.73, Prob > chi2: 0, Pseudo R2: 0.3, Log pseudolikelihood: -1591.929, Cluster adjustments: 157. Model 1b: N: 4,798, Wald chi2(7): 59.15, Prob > chi2: 0, Pseudo R2: 0.3983, Log pseudolikelihood: -1151.7899, Cluster adjustments: 152. Model 1c: N: 5,692, Wald chi2(14): 76.12, Prob > chi2: 0, Pseudo R2: 0.4471, Log pseudolikelihood: -1221.1816, Cluster adjustments: 155. Model 1d: N: 4,606, Wald chi2(15): 99.92, Prob > chi2: 0, Pseudo R2: 0.4825, Log pseudolikelihood: -962.57728, Cluster adjustments: 150.

$^* p < .10$ $^{**} p < .05$ $^{***} p < .01$.

TABLE 4.2B
Predicting the Diffusion of Nuclear Weapons in Stages:
Nuclear Weapons Acquisition

	No latent nuclear variable, no controls	Latent nuclear variable, no controls	No latent nuclear variable, controls	Latent nuclear variable, controls
	Model 2a	Model 2b	Model 2c	Model 2d
GDP per	0.0002	0.0002	0.0003	0.001
capita	(0.0001)**	(0.0001)	(0.0001)**	(0.0002)***
GDP squared	−1.54E−08	−1.75E−08	−2.04E−08	−3.95E−08
	(5.12E−09)***	(6.73E−09)***	(5.59E−09)***	(1.12E−08)***
Industrial	12.581	7.052	14.520	18.737
capacity	(7.606)*	(7.233)	(8.284)*	(9.754)*
Latent nuclear		0.583		0.842
capacity		(0.256)**		(0.366)**
Military	1.15E−07	1.28E−07	1.02E−07	1.39E−07
spending	(2.25E−08)***	(2.48E−08)***	(1.19E−08)***	(2.69E−08)***
Diffusion	0.437	0.554	1.386	2.419
	(0.324)	(0.517)	(0.657)**	(0.815)***
Organizational	0.017	0.024	0.021	0.036
age	(0.007)**	(0.012)**	(0.008)***	(0.013)***
Military			0.0001	0.000
personnel			(0.0002)	(0.0003)
Polity			0.118	0.101
			(0.026)***	(0.033)***
Nuclear ally			−2.817	−3.077
			(0.706)***	(0.613)***
Nuclear threat			1.424	1.772
			(0.604)**	(0.674)***
Dispute			0.285	0.388
propensity			(0.060)***	(0.090)***
Security			0.070	−0.754
threat			(0.618)	(0.661)
Global			1.499	0.973
satisfaction			(1.463)	(2.180)
Regional			−0.894	−0.867
satisfaction			(0.481)*	(0.688)
Constant	−4.973	−9.089	−11.613	−21.580
	(1.191)***	(2.220)***	(2.832)***	(5.088)***

Notes: Robust standard errors in parentheses. Model 2a: N: 5,971, Wald chi2(6): 57.12, Prob > chi2: 0, Pseudo R2: 0.612, Log pseudolikelihood: −471.74321, Cluster adjustments: 157. Model 2b: N: 4,798, Wald chi2(7): 52.23, Prob > chi2: 0, Pseudo R2: 0.6855, Log pseudolikelihood: −304.2704, Cluster adjustments: 152. Model 2c: N: 5,692, Wald chi2(14): 644.89, Prob > chi2: 0, Pseudo R2: 0.7896, Log pseudolikelihood: −252.60844, Cluster adjustments: 155. Model 2d: N: 4,606, Wald chi2(15): 278.93, Prob > chi2: 0, Pseudo R2: 0.8483, Log pseudolikelihood: −145.22624, Cluster adjustments: 150.

* $p < .10$ ** $p < .05$ *** $p < .01$.

TABLE 4.2C
Predicting the Diffusion of Nuclear Weapons in Stages: Nuclear Weapons
Acquisition Censored on Nuclear Weapons Program

	No latent nuclear variable, no controls	Latent nuclear variable, no controls	No latent nuclear variable, controls	Latent nuclear variable, controls
	Model 3a	Model 3b	Model 3c	Model 3d
GDP per capita	0.0001 (0.0002)	0.0002 (0.0001)	0.001 (0.0004)**	0.001 (0.0004)**
GDP squared	−1.67E−09 (−1.08E−08)	−1.04E−08 (1.06E−08)	−3.69E−08 (1.96E−08)*	−4.54E−08 (1.83E−08)**
Industrial capacity	35.552 (14.186)**	27.733 (15.091)*	42.857 (12.853)***	43.268 (14.233)***
Latent nuclear capacity		0.684 (0.352)*		0.243 (0.186)
Military spending	1.34E−07 (5.36E−08)**	1.51E−07 (5.25E−08)***	5.40E−08 (2.68E−08)**	3.06E−08 (2.73E−08)
Diffusion	2.229 (0.691)***	2.290 (0.730)***	4.262 (0.891)***	5.157 (1.167)***
Organizational age	0.006 (0.009)	0.019 (0.012)	0.013 (0.012)	0.018 (0.012)
Military personnel			0.0001 (0.0003)	0.001 (0.0003)*
Polity			0.106 (0.055)*	0.125 (0.058)**
Nuclear ally			−1.852 (0.451)***	−2.016 (0.428)***
Nuclear threat			3.506 (1.926)*	2.706 (1.601)*
Dispute propensity			0.414 (0.079)***	0.442 (0.100)***
Security threat			−0.086 (0.517)	−0.319 (0.613)
Global satisfaction			2.695 (1.760)	1.623 (1.921)
Regional satisfaction			−1.049 (0.707)	−1.444 (0.798)*
Constant	−11.072 (2.808)***	−16.263 (2.661)***	−25.201 (4.269)***	−30.060 (5.414)***

Notes: Robust standard errors in parentheses. Model 3a: N: 759, Wald chi2(6): 49.74, Prob > chi2: 0, Pseudo R2: 0.6507, Log pseudolikelihood: −179.16557, Cluster adjustments: 23. Model 3b: N: 656, Wald chi2(7): 76.87, Prob > chi2: 0, Pseudo R2: 0.6891, Log pseudolikelihood: −134.77476, Cluster adjustments: 23. Model 3c: N: 745, Wald chi2(14): 447.41, Prob > chi2: 0, Pseudo R2: 0.8309, Log pseudolikelihood: −85.498258, Cluster adjustments: 23. Model 3d: N: 642, Wald chi2(15): 399.84, Prob > chi2: 0, Pseudo R2: 0.8393, Log pseudolikelihood: −68.592364, Cluster adjustments: 23.

$* p < .10 ** p < .05 *** p < .01.$

of the relationship between the independent variables and two stages of the nuclear weapons acquisition process: the initiation of a nuclear weapons program and the actual acquisition of nuclear weapons. Each model has Huber-White robust standard errors and clusters on the individual countries involved to correct for bias from multiple observations from the same country. Model 1 is a probit model of the decision to pursue nuclear weapons while the dependent variable in model 2 is the acquisition of nuclear weapons. The dependent variable in model 3 is also nuclear weapons acquisition, but the regression is censored on the basis of pursuing nuclear weapons, meaning it only includes countries that attempted to pursue the acquisition of nuclear weapons.[30]

All three models show a significant relationship between adoption-capacity variables and the adoption of nuclear weapons, although their impact changes in the shift from initiating a nuclear weapons program to actually building nuclear weapons. In model 1, focusing on the decision of a country to begin a nuclear weapons program, GDP squared is negatively related to the decision to begin a nuclear weapons program, as predicted, while GDP per capita is positive. The results suggest that basic economic development makes it more likely a country will pursue nuclear weapons, but only up to a certain level, at which the effect reverses. First differences on GDP per capita and GDP squared show that the simultaneous move from the mean value of both variables to the maximum decreases the probability a nuclear weapons program begins by .059 to almost 0. Yet the importance of this effect is limited by the varying significance of the GDP per capita variable depending on the model specification. The latent nuclear weapons capability variable is positively and significantly related to the probability of beginning a nuclear weapons program, with a first difference of .23, as is military spending (the first difference is generated using Clarify [King, Tomz, and Wittenberg 2000]). The first difference for military spending reveals an increase from .03 to over .9 as military spending shifts from the minimum to the maximum. Interestingly, the raw industrial-capacity variable is not significant. In combination, these results suggest that adoption-capacity constraints on the financial side do play an important role in influencing decisions about pursuing nuclear weapons. The organizational age variable is mildly significant and positive—the expected direction for an innovation requiring low levels of organizational capital for adoption. The first differences for it are interesting: an increase from the minimum value to the 90th percentile yields an increase of .09 in the probability of a nuclear weapons program, or 9 percent. An increase from the 90th percentile to the maximum increases the probability an astonishing .7 alone. This is probably the result of the first-moving United States and early adopters like the Soviet Union and Great Britain, which avoided having their organizational age "reset" as a function of their victories in World War II.

[30] This is consistent with Jo and Gartzke 2007.

Turning to the models of the adoption of nuclear weapons, models 2 and 3, the adoption-capacity variables demonstrate a powerful and significant relationship between industrial capacity and the probability that a state develops nuclear weapons. First differences show an increase from .07 for states with a minimal level of industrial capacity to over .3 for states in the 70th percentile of industrial capacity. The number then jumps above .9 as the variable moves toward its maximum, revealing that those states with more advanced industrial capacities are substantively much more likely to adopt. As described above, model 3 is censored—it has over thirty-five hundred fewer observations than the others—because it only includes observations where a state in a given year has a nuclear weapons program. The GDP per capita and GDP squared variables have similar values to the program initiation equation, though their substantive effect is more clearly u-shaped. At their minimums, the probability of adoption is .03, but the probability jumps to .6 when the variables are at their means before declining again to .03 at their maximums. Interestingly, latent nuclear capacity (when included in the models) and military spending are always not statistically significant despite their strong relationship to the initiation of a nuclear weapons program. The difference between the program initiation and adoption models suggests that elements of adoption capacity matter at different points in the nuclear weapons development process. High-spending states with a basic level of economic development and latent nuclear capacity are more likely to initiate a nuclear weapons program. When it comes to adoption, though, strong industrial economies are significantly more likely to actually acquire nuclear weapons—a trend that may change over time. Alternatively, the positive and significant diffusion variable in the adoption equations shows, as predicted, that the adoption-capacity requirements over time for nuclear weapons may have declined given the diffusion of information and expertise about nuclear weapons, though they still remain high. The results are also consistent using a selection model.

The control variables perform in expected ways, further increasing our confidence in their inclusion, though the results are presented without them as well. Security challenges appear less related to the initiation of nuclear weapons program than the decision to follow through and build a nuclear weapon—a finding consistent with theory. The costs of beginning a nuclear weapons program, both physical and in the international realm, are much lower than taking the final step (even though there are potential benefits too, as I discuss below) and actually building a nuclear weapon. Therefore, being allied with a nuclear power or facing a nuclear-armed adversary is not significantly related to creating a nuclear weapons program, but countries with nuclear armed allies are significantly less likely to follow through and build a nuclear bomb, while states facing a nuclear-armed adversary are much more likely to adopt. This statistical result fits with the predictions of adoption-capacity theory. U.S. allies like Australia and South Korea began nuclear weapons programs during the cold war but then abandoned them. While their respective alliances with the United States were not solely responsible for those decisions, the United States' extended deterrence promises undoubtedly helped these states to pass the buck on spending to acquire nuclear weapons, instead choosing to pay the "cost" of a lesser alliance role in exchange for protection.[31]

These statistical results provide crucial validation for diffusion-related hypotheses in the nuclear case. The increasing importance of financial intensity as states get closer to acquiring nuclear weapons explains why so many states have moved through the early stages of nuclear weapons programs only to fall back on balancing and bandwagoning strategies. Questions of shifts to neutrality and countering as dominant strategies are dealt with below because they more appropriately fit into the discussion of the impact of nuclear weapons on global politics.

Nuclear Impact Research Design

While the appropriate model for testing the spread of nuclear weapons was a country year setup that evaluated the actions of each country in every given year, testing many of the hypotheses regarding the impact of nuclear weapons on international politics requires a more interactive approach. The main nuclear weapons impact models below use the MIDs 3.0 data set (Ghosn and Bennett 2003).[32] The data is set up in directed dyads for nearly all of the tests below, meaning it describes the interaction of pairs of states, where side A refers to the challenger, the state that originally initiated the dispute by threatening, deploying, or using military force against another state, and side B refers

---

[31] Finally, as discussed above, the presence of the NPT, an international treaty regulating the spread of nuclear weapons, may act as a potential inhibitor of diffusion. A final robustness test checks the potential relevance of the NPT. A dichotomous *NPT* variable measures whether or not a given state in a country year has ratified the NPT. Given that nuclear taboo and international norms arguments heavily rely on the weight of treaties as a factor that influences national preferences, an *NPT System* variable is created to measure the percentage of states in the system in a given country year that have ratified the NPT (Finnemore and Sikkink 1998; Tannenwald and Price 1996). The NPT system variable is drawn from Jo and Gartzke 2007, 175. Both adherence to the NPT by an individual country and the level of NPT adherence in the international system may play a weak but significant role discouraging states from converting their nuclear weapons programs into arsenals. Interestingly, there is a positive correlation between global NPT adherence and nuclear weapons programs, perhaps because states know any punishments from the International Atomic Energy Agency or UN Security Council will be slow at best. But at the nuclear weapons acquisition level, the NPT appears to discourage proliferation. Nevertheless, the NPT variables are also highly correlated with other independent variables, raising the question of whether or not the NPT variable or, say, the diffusion timing variables are really "doing the work" in driving the results. This is an intriguing avenue for future research.

[32] MIDs version 3.02 adds 300 MIDs (500 MID dyads), from 1993 to 2001, covering instances where one state threatens, displays, or uses military force against another state.

to the defender.[33] Since the hypotheses often depend on the relative capabilities of states in the international system, a directed dyads approach is necessary to capture the specific effect of nuclear weapons on militarized disputes.

Nuclear Impact Dependent Variables

The central predictions for the impact of nuclear weapons on the international security environment relate to their reinforcement of the gap between major powers and other states, and the success of those states in militarized situations. The dependent variable for measuring success, building on prior work by Kenneth Schultz (1999) on coercive bargaining, is whether or not a defender reciprocates an initial militarized action by a challenger or backs down.[34] The decision to respond to a militarized challenge with a militarized action is a significant escalatory step—the joining of the dispute by the defender—in the dispute process.

Testing the predictions that nuclearized states should experience a broader range of militarized actions with a larger number of states over greater distances is conducted by measuring MID intensity, the average number of MID entrances in a five-year period, and the geographic distances of MID participants between nuclear and nonnuclear states.[35] The major power dropout hypothesis is testable by using COW's defined major power entries and exits as the dependent variable. A state that experiences a major power dropout is coded a 1 for the two years prior and two years following the dropout as well as the dropout year, and a 0 at all other times.

Nuclear Impact Independent Variables

The key independent variable of interest for these tests is whether or not each side in a dyad in a given year possesses nuclear weapons—*Side A Nuclear* and *Side B Nuclear*—the dependent variable from the nuclear diffusion tests. Control variables designed to isolate the specific impact of nuclear weapons on international politics include many variables already described above for the

---

[33] For an explanation of the merits of utilizing dyad years as the unit of analysis, see Bennett and Stam 2004; Gleditsch and Hegre 1997.

[34] The data are drawn from the MIDs dispute dyads data set. The MIDs data set codes actions by each side on a 1 to 5 scale: 1 is for no militarized action, 2 is for the threat to use force, 3 is for the display of force, 4 is for the use of force, and 5 is for war. Once the challenger initiates a militarized dispute, the defender essentially has two choices: respond with their own threat, display, or use of force (2–5), or back down (1) and not respond with any militarized actions. If the defender responds with a militarized action, meaning an action coded by the MIDs data set as a 2 or above, it is coded 1 in the data set. If the defender is coded as not responding militarily, the dependent variable is coded 0.

[35] The data on geographic distances are drawn from EUGene (Bennett and Stam 2000).

nuclear diffusion model. A measure of the conventional balance of forces in each dyad year was also included, *Conventional Balance of Forces*. Using relative national power scores (combining measures of economic, demographic, and military power) as coded by the COW data set, the balance of power control measures the gap between the relative power of side A and the relative power of side B.[36]

Another control evaluates the relative degree of preexisting animosity in a dyad, which might predispose it to conflict regardless of nuclear weapons. Based on the "satisfaction" data described above, *Dyadic Satisfaction*, generated using the Bueno de Mesquita Tau-B scheme, measures the satisfaction of both states in the dyad based on their alliance portfolios. The variable, as modified by Bennett and Stam, runs from 0 to 1, as did the regional and global satisfaction variables in the diffusion tests.

Finally, the dyadic tests also include controls for the particular issue involved in the dispute with three variables, *Territory*, *Regime/Government Change*, and *Policy*. Research by Schultz (1999) as well as Bruce Bueno de Mesquita, James Morrow, and Ethan Zorick (1997) shows that the issue under dispute influences the way actors behave. Some issues, like territorial control, are often valued more than simple policy changes. The MID data for each dispute include a four-tiered variable for both side A and side B measuring whether each state has revisionist demands, and the type of demand: the issue categories are territory, regime/governmental change, policy, and other.[37] Dichotomous variables are created for each issue category. This will help control for the potentially spurious impact of nuclear weapons in cases where perceptions by the challenging state about the importance of the issue might really be driving behavior.

Nuclear Impact Results

First-Mover Advantage

Evidence from the early cold war suggests that the U.S. nuclear monopoly played a limited but important role in boosting the relative power of the United States in the aftermath of World War II, highlighting an advantage for the first-moving state.[38] John Lewis Gaddis actually shows that shortly after Hiroshima, the Truman administration had already decided that maintaining

---

[36] The equation used to produce the relative power variable was cap_1/(cap_1 + cap_2). Using an alternative conventional capability measure (military spending and personnel) did not significantly change the results.

[37] The "other" revisionist issue variable was dropped due to extremely high collinearity.

[38] While the original plan was to statistically test the gain to the United States from the nuclear monopoly, according to MIDs 3.02, the United States was only involved in three militarized

a long-term monopoly on nuclear weapons was untenable, making nuclear monopoly an ineffective strategic lever. The multinational nature of the Manhattan Project and the information being published by U.S. scientists on their efforts also seemed to suggest in late 1945 that efforts to fully control nuclear weapons would not succeed (Gaddis 1997, 90–91).

Nevertheless, budget constraints and a perception of conventional weakness in comparison with the Soviet Union soon turned the Truman administration toward utilizing its nuclear monopoly for a strategic edge. The imposed $15 billion budget ceiling made nuclear weapons a potentially more efficient means of ensuring U.S. defense than a conventional buildup (Gaddis 1997, 91–92; Williamson and Reardon 1993).

The discussion of preventive war throughout the U.S. defense establishment in the early cold war period underscores the relevance of the nuclear monopoly for U.S. strategy. General Leslie Groves, head of the Manhattan Project, argued in a 1946 memorandum that the nuclear monopoly was so essential that the United States should consider preemptively striking the nuclear facilities of any country before it gained atomic capabilities (Trachtenberg 1988, 5). Statements by public intellectuals like Bertrand Russell and reporters such as William Laurence also demonstrate serious concern about the consequences for the United States when its nuclear monopoly eroded (7–8).

Soviet leader Stalin also clearly perceived nuclear weapons as a difference maker in international politics—at a minimum as a psychological weapon of intimidation. While Vyacheslav Molotov publicly asserted the irrelevance of nuclear weapons at the London Conference in September 1945, Stalin claimed that "Hiroshima has shaken the whole world" (Gaddis 1997, 92). Stalin believed that the United States and Great Britain could leverage the atomic monopoly to ensure a favorable outcome in negotiations over the future of Europe, mandating a crash Soviet atomic weapons program and a counterstrategy to prevent U.S. coercive success (92). In his study of Stalin's early response to nuclear weapons, David Holloway (1994, 258–63) contends that nuclear weapons forced Soviet restraint in the use of force, for fear of escalation, but also encouraged Soviet recalcitrance for fear of appearing weak.

One example of how the nuclear monopoly influenced U.S. foreign policy is the way that the United States responded during the first Berlin airlift crisis (1948–49). According to a memo sent by Foy Kohler, the U.S. chargé d'affaires in the Soviet Union, U.S. possession of the atomic bomb uniquely allowed the United States to stand firm in the early days of the dispute. Soviet apprehension about fighting a nuclear-armed adversary, especially in a war over Berlin, a

---

disputes that took place entirely during the period of the U.S. nuclear monopoly. This made the sample too small to statistically test.

territory some thought Stalin did not consider worth a conventional or nuclear war, influenced the U.S. decision.[39]

So the United States did acquire a clear first-mover advantage from its nuclear monopoly. The nuclear monopoly enabled the United States to survive an era of conventional weakness vis-à-vis the Soviet Union in the immediate aftermath of World War II. Confidence in the guarantor of U.S. survival and threat to others provided by the atomic bomb allowed U.S. leaders like Harry S. Truman, whether by design or not, to resist the temptation to cede Europe to the Soviet Union and retreat across the oceans back to the United States. While the lessons of history from the post–World War I period, the cost of isolationism, and the danger of appeasement may have formed the intellectual core of the new U.S. approach, fear of the U.S. nuclear monopoly helped hold Stalin in check. Belief in the power of the atomic bomb enabled the United States to supply Berlin and advocate for the formation of NATO.

Nuclear weapons appear to offer late-mover advantages as well, although it is somewhat unclear whether this is just a demonstration of the importance of nuclear weapons in general. Table 4.3 compares the GDP of nuclear weapons adopters in the year they acquired nuclear weapons to the GDP of the United States in 1945, the year of the first nuclear explosions.[40] The evidence above suggests that the global spread of tacit knowledge about the production of nuclear weapons, from scientific publications, declassified documents, and scientists willing to sell their knowledge to the highest bidder, has lowered the relative costs associated with acquiring nuclear weapons. Yet the cost of adoption is simply too high for late adopters to acquire the most technologically advanced weapons or a large number of weapons, especially in the short term. There is certainly a financial advantage to late adoption that also provides new nuclear states with enhanced coercive power—just not a military effectiveness advantage—that would mean late adopters have a greater opportunity to build more and more modern nuclear weapons than early adopters.

The idea that late adopters more efficiently implement innovations than early ones by learning from past experience, though true in the case of the United States' experimenting on dead-end projects that ultimately saved others from having to do the same, is less relevant to the nuclear weapons case.

---

[39] Minutes of the 286th Policy Planning Staff Meeting, September 28, 1948, U.S. Department of State 1973, 1196; Kohler (the chargé in the Soviet Union) to George Marshall (Secretary of State), September 28, 1948, U.S. Department of State 1974, 920. The relative strength of the U.S. atomic arsenal allowed the United States to oppose Soviet actions (Botti 1996; Druks 1967, 168–69; Halperin 1987, 7). Interestingly, Gaddis views Harry S. Truman's decision to deploy B-29s to Europe in the first Berlin crisis as less of a nuclear signal than it is traditionally interpreted to be. Gaddis (1997, 91) argues that since the planes were not equipped with nuclear bombs, a fact the Soviets could observe, their deterrence value was questionable.

[40] Instability in the GDP data for 1944, probably due to World War II, caused the use of 1945 instead. Substituting 1944 does not change the results.

TABLE 4.3
Relative Per Capita GDP of Nuclear Weapons Adopters
(Compared to United States in 1945)

Country	Year	Real per capita GDP (purchasing power parity, 1996 U.S. dollars)	Comparison with United States in 1945
Soviet Union	1949	3,359	31.69%
United Kingdom	1952	7,578	71.50%
France	1960	7,825	73.83%
China	1964	714	6.73%
Israel	1967	6,515	61.48%
India (1)	1974	1,019	9.62%
South Africa	1979	7,599	71.70%
Pakistan	1987	1,580	14.90%
India (2)	1988	1,561	14.72%

Source: Data drawn from the Penn World tables version 6.2 (Heston, Summer, and Aten 2006; Singh and Way 2004).

The cost of developing a nuclear arsenal is so high that even if the relative financial and organizational cost of developing the first nuclear weapon has declined, the costs of developing stable delivery mechanisms, large numbers of weapons, and reliable warheads are still large.[41]

Statistical Evidence about the Coercive Success of Nuclear States

Given the dichotomous nature of the dependent variable described above, whether or not a militarized dispute initiation is reciprocated, the most appropriate statistical model is logistic regression. The data set, drawn from EUGene, includes all militarized disputes from 1945 to 2001, a total of 1,726.[42] The model uses Huber-White robust standard errors to hedge against potential heteroscedasticity. The possibility of fixed time and country effects was corrected with the addition of peace-year controls (Beck, Katz, and Tucker 1998).[43] Table 4.4 shows the relationship between nuclear weapons

[41] This could just be because these states do not need big arsenals. Then again, most of the states that have recently tried to acquire nuclear weapons are not major industrialized ones. It would be interesting to see how quickly a state like Germany or Japan, if it acquired nuclear weapons, could move beyond the first few weapons, and develop diversified delivery mechanisms and a larger arsenal.

[42] To test for the possibility that nuclear weapons are a proxy for something that influences the probability of selection into disputes but not escalation, the author also ran a Heckman-style bivariate probit model. MID initiation was selection into the model, and reciprocation was the secondary dependent variable. The results confirm the importance of nuclear weapons at both stages.

[43] A peace-year count variable (time since the last MID) also had no impact on the results.

TABLE 4.4

Statistical Relationship between Nuclear Weapons Possession and Coercive Success in Directed Dyads, 1946–2001

	No controls (logit)		Full specification (logit)		Conditional logit 1 (nuclear ID grouping)		Conditional logit 2 (dyadic grouping)	
	Coefficient	Robust standard error	Coefficient	Robust standard error	Coefficient	Robust standard error	Coefficient	Robust standard error
Side A nuclear	−0.487***	0.125	−0.321**	0.161	−0.816***	0.158	−0.840*	0.437
Side B nuclear	−0.118	0.136	0.107	0.166	−0.326*	0.193	−0.118	0.46
Conventional balance of forces			0.171	0.206	0.218	0.517	−0.657	1.429
Dyadic satisfaction			0.192	0.243	0.418***	0.145	0.124	0.835
Side A democracy			−0.265*	0.139	−0.418***	0.088	0.49	0.333
Side B democracy			−0.243**	0.117	−0.392**	0.151	0.209	0.424
Peace years 1	0.001***	0	0.001**	0	0.001***	0	−0.001	0.001
Peace years 2	0.000**	0	0.000**	0	0.000***	0	0.001	0.001
Peace years 3	0	0	0	0	0.000***	0	−0.001**	0
Territory			0.465***	0.154	0.443*	0.227	0.12	0.265
Regime/government			0.185	0.268	0.215	0.435	0.014	0.439
Policy			−1.220***	0.132	−1.167***	0.095	−1.015***	0.202
Constant	0.200***	0.065	0.482**	0.214				
N	1,706		1,669		1,669		1,028	
Wald chi2	5,69.88		12,247.65		1,..		12,53.3	
Prob > chi2	0		0		.		0	
Pseudo R2	0.033		0.124		0.135		0.072	
Log pseudo likelihood	−1,140.485		−1,011.274		−991.673		−398.687	

Note: The first fixed effects model was run with nuclear countries as the group identification variable. The second fixed effects model was run with unique dyadic combinations as the group identification variable.

* $p < .10$ ** $p < .05$ *** $p < .01$.

possession and MID reciprocation, displaying a simple model without any control variables and a larger model including relevant controls. In all of the models, the possession of nuclear weapons by side A, the challenging side in the dispute, makes the probability of reciprocation by side B significantly less likely, meaning nuclear weapons provide a substantial boost to the probability of "success" for side A. These results highlight the importance of nuclear weapons in bringing militarized dispute success to nation-states.

The fixed-effects conditional logit models control for the way interactions between individual states over time, as opposed to in single disputes, may distort the results. In the fixed-effects model the nuclear weapons variable is also significant, revealing its relevance in comparison with other factors like the conventional balance of power or dyadic satisfaction.[44]

These results make sense given the prediction that nuclear weapons should influence the calculations of states about the relative costs of conflict. Even if a state never makes an explicit nuclear threat, the mere presence of a nuclear weapon may exert a powerful coercive role in low-level militarized disputes. For example, when U.S. troops deployed to Lebanon in 1958 in response to a coup in Iraq, U.S. President Dwight Eisenhower's decision to place nuclear-armed bombers on alert and forward deploy tanker planes arguably played a role in demonstrating credible U.S. resolve to the Soviet Union. The Soviet Union did not provide substantial support to Egypt in the crisis even though the United States never made even an implicit nuclear threat. Khrushchev reportedly told Gamal Abdel Nasser that the Soviet Union was not ready for World War III (Betts 1987, 66–68). This explains Betts's statement that there is "a whiff of blackmail inherent in standing capabilities and doctrines" (6).

Figure 4.2 shows the substantive impact of nuclear weapons possession on the probability of MID reciprocation in the fully specified logit model above given different configurations of nuclear capabilities on sides A and B. These results highlight how variations in nuclear capabilities within a dyad can influence militarized interactions, though the impact is limited. The highest probability of MID reciprocation comes when only side B has nuclear weapons; a nuclear advantage gives defenders confidence they can control the situation and gain advantages if escalation occurs. The lowest probability of reciprocation—11 percent lower—occurs when side A has an asymmetrical nuclear advantage, meaning side B does not want to risk escalation against a nuclear-armed state. Interestingly, the probability of reciprocation for non-nuclear dyads is slightly higher than that of nuclear dyads. Consistent with the theorizing above and similar to the predictions of scholars like Waltz (1995), in some particular instances the possession of nuclear weapons by

---

[44] Additional models available on request from the author include robustness checks for different specifications of the regime type variables and the nuclear weapons variables.

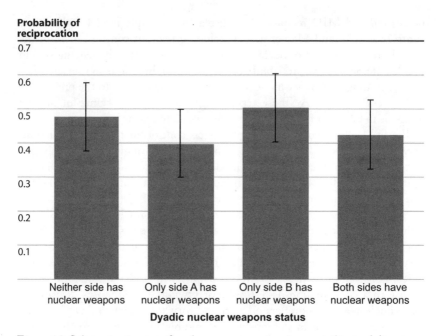

Figure 4.2. Substantive impact of nuclear weapons possession on militarized dispute reciprocation.

both sides in a dispute may cause restraint because both sides fear the ultimate consequence of nuclear escalation.[45]

An alternative explanation for the results is that more experienced nuclear states have larger arsenals, and that arsenal size and sophistication drive the behavior of nuclear states. A control run that added a variable measuring the nuclear arsenal size of each nuclear state in a given year was estimated. The results did not substantially change. This is likely because only a small fraction of total disputes, 82 of 1,726, come from interactions between two nuclear states. Therefore, while the nuclear balance of power is obviously a critical factor governing the behavior of nuclear states in militarized disputes against each other, the theory is more focused on asymmetrical nuclear interactions.[46]

---

[45] Other statistical work, however, shows that this finding may not apply to new nuclear proliferators. New nuclear states are especially likely to initiate disputes and reciprocate challenges regardless of the nuclear status of the potential opponent. See Horowitz 2009.

[46] A nuclear second-strike variable was not added because it is difficult to measure whether or not a nuclear state has second-strike capabilities, as policy debates regarding U.S. and Soviet capabilities demonstrate, and it does not apply for many of the nuclear states.

TABLE 4.5
Relative MID Intensity Levels—Nuclear versus Nonnuclear, 1945–2002

	Nonnuclear	Nuclear
MID fatality level 0	812	426
MID fatality level 1	128	32
MID fatality level 2	57	13
MID fatality level 3	24	7
MID fatality level 4	12	2
MID fatality level 5	3	0
MID fatality level 6	23	13
Total	1,059	493
Proportion of total cases in category with fatalities	0.2332	0.1359
% with fatalities	23.32	13.59
% with 500 + casualties	10.53	19.40
% with 1K + casualties	9.31	19.40

MID Frequency, Intensity, and Geography

Table 4.5 shows the relative intensity levels for MIDs involving nuclear and nonnuclear powers in the post–World War II era. One representation of intensity is casualties, which highlight the commitment to the dispute by both sides. Only 9 to 10 percent of those MIDs that escalate to involve fatalities with nonnuclear initiators have over five hundred casualties. In contrast, almost 19 percent of those MIDs involving nuclear-armed initiators have escalated to involve over five hundred total casualties.

The relationship between acquiring nuclear weapons and the geographic scope of MIDs points to the broader geographic reach of nuclear states, providing additional evidence in favor of the military interactions hypotheses of adoption-capacity theory. Each MID is ranked based on the distance between capitals, and being placed in a higher percentile signifies a greater distance between capitals, relative to the distances between capitals for other MIDs.[47] Table 4.6 shows that the average dispute distances for nuclear powers are not only longer at every percentage interval; the median of zero for nonnuclear states illustrates that the higher numbers toward the maximum skew the nonnuclear distribution. Thus, it appears that nuclear powers more regularly engage in disputes over greater distances to an extent even greater

[47] Geography data drawn from Bennett and Stam's EUGene software program (which itself draws from the 1993 COW contiguity data, among other places, for the geography data) and the distances are distances between capitals calculated using the "great circle" distance formula (Bennett and Stam 2000).

TABLE 4.6

The Geography of Militarized Disputes and Dispute Propensity of Militarized
States—Nuclear versus Nonnuclear

	Geography		Dispute propensity	
	Side A nuclear	Side A nonnuclear	Nuclear	Nonnuclear
Observations	294	1,247	309	6,671
Minimum	0	0	0	0
10th percentile	0	0	0.6	0
25th percentile	0	0	1.4	0
50th percentile	1044	0	2.6	0.2
Mean	1,986.74	1,056.31	3.19	0.53
75th percentile	4,291	1,140	4.2	0.6
90th percentile	5,814	4,060	7	1.4
Maximum	11,809	11,718	12.6	17.75

than the initial evidence might reveal.[48] Calculating the five-year moving average of militarized dispute involvement for all states during the period similarly demonstrates the higher propensity of nuclear states to engage in militarized disputes in the first place.[49] It is certainly true that there is a correlation between relative power, in general, and the states that have acquired nuclear weapons. The major powers are more likely in general to have more expansive geographic interests. As the tables above show, though, even controlling for the relative power of a state, nuclear weapons have an independent impact.

Major Power Dropout/Power Transition

There is little evidence for or against the major power dropout and power transition hypotheses. According to the Correlates of War Project (2007) update to the list of major powers from 1816 to 2004, no state that ended World War II as a major power has lost that designation in the interim. Meanwhile, China became a major power in 1950, while Germany and Japan both became major

[48] This only focuses on side A since side A is always the dispute initiator in this dyadic setup and only side A chooses the distance at which it will engage in a dispute.

[49] The results are available on request. While the bivariate and simple statistical nature of this test makes the results tentative, in combination with the other results the evidence forms a clear pattern. The dispute propensity data are set up in country-year format, while the geographic distance data are arranged in dyadic format, which explains why the geography labels include a side A. There are many more observations for the dispute propensity data because they are all country years after 1945, while the geography data are only for militarized disputes after 1945. Missing data explain the difference in cases between the geography and fatality tables.

powers in 1991. Of the current major powers, the only ones without nuclear weapons are Germany and Japan. Both exist under nuclear security umbrellas, with Germany's arguably originating from multiple locations: the United States, Great Britain, and France. Japan is also widely known to be only a short time away from adopting nuclear weapons whenever it decides to do so, given its extensive and highly developed civilian nuclear power industry, though it would take at least three to five years to develop a nuclear force according to a recent Japanese study (Green and Furukawa 2000; Mack 1996; Sankei Shimbun 2007).

Conclusion

The evidence shows that relative financial intensity levels powerfully predict both the ability of a state to initiate a nuclear weapons program, and whether or not it will eventually acquire nuclear weapons. In contrast to research on "risky" states, which tends to focus on what determines the desire to acquire nuclear weapons, adoption-capacity theory demonstrates that the high level of financial intensity required for adoption drives the bigger nuclear picture in part. But the decreasing financial intensity required for adoption over time has opened the door to new adopters. Unlike with a battleship or aircraft carrier, a state can invest gradually over time in a nuclear program and the end result will still be highly relevant for international politics. As time goes by and tacit knowledge leaks even more, it will be increasingly difficult to maintain constraints on nuclear proliferation. The inherent attraction of nuclear weapons as a difference maker in international politics also means that for states to forgo the nuclear option and choose another strategic path, they need to receive other benefits, possibly in the form of alliances. Nuclear-related alliances contain benefits for both sides. The nonnuclear state receives protection: a guarantee that the nuclear state will protect it in a crisis, especially against another nuclear state. The nuclear state, in turn, guarantees that one fewer state will acquire the atomic bomb.

The impact of nuclear weapons on international politics is also consistent with adoption-capacity theory. As a financially intense innovation with low organizational barriers to entry, nuclear weapons have traditionally widened the gap between the major powers and other states in the international system by setting up a global power litmus test. The possession of nuclear weapons is, unsurprisingly, significantly related to success in militarized disputes. But the cumulative nature of nuclear investments mitigates the power-balance-reinforcing nature of the nuclear weapons innovation. Because even rudimentary nuclear weapons are considered powerful, states can invest over a period of a decade or two in a nuclear weapons program and reap the benefits at the end, explaining the spread of nuclear weapons beyond what would otherwise be a small club.

Chapter 5

BATTLEFLEET WARFARE

On October 21, 1805, Admiral Horatio Nelson and his British fleet wiped out a combined French and Spanish fleet at Trafalgar, ushering in decades of British naval superiority. In 1828, following a victory over the Ottoman Navy at Navarino in 1827 and conscious of their global leadership in sea power, the lords of the British Admiralty issued a regulation discouraging the development of steam-powered boats with the logic that any naval technological innovations would only threaten Britain's naval predominance. These regulations controlled the British Navy for thirty years; in 1851 the British Navy still looked like it had for much of the last three centuries, dominated by ships of the line. After the introduction of the great French ironclad, *La Gloire*, in 1858, and reports of the epic 1862 U.S. naval ironclad clash between the *Monitor* and the *Merrimack*, however, the British rapidly shifted their ship production to iron and then steel warships propelled by steam instead of sail. But most other elements of naval strategy remained the same; doctrine, education, organizational structure, promotions, recruitment, and training all looked almost exactly the same in the 1880s as they had at Trafalgar.

Yet in the short period of time that followed, in the run-up before World War I, the greatest naval power in the world overhauled everything from its doctrine and training to force structure. This shift, launched in 1906 with the HMS *Dreadnought* as its public face, represents one of the only cases of a dominant global military power self-consciously reshaping the core competencies that had guaranteed its success without ever losing a battle. This shift, sometimes called the Fisher Revolution after the fiery leadership of First Sea Lord John Fisher, but called the battlefleet warfare innovation in this book, helps to define why understanding the diffusion of major military innovations is essential to grasping the development, spread, and use of military power in general.

The introduction of the HMS *Dreadnought* and the organizational changes accompanying battlefleet warfare drove substantial changes in the way that the major naval powers designed and planned for warfare. On the technological side alone, navies around the world froze their construction plans for over a year as they attempted to digest what the *Dreadnought*, with its uniform caliber guns and fast propulsion system, would mean for naval warfare. The rising cost per unit of dreadnought-style ships also meant that the total number of capital ships possessed by a navy declined even as the relative power of the navy increased due to the greater capabilities of the ships themselves. Initially this aided the British Navy by helping it save money. But as the number of ships necessary to be a great naval power declined, it also opened the door for rising naval powers, most notably Germany, to more effectively challenge British naval supremacy than might have been possible in the previous period.

While the British Navy did not have to lose a battle at sea to change, those in favor of transformation did have to win a bureaucratic struggle for control over the Admiralty, which controlled funds and decision making. In the end, it was the organizational changes instituted by the British naval leadership that allowed Britain to effectively harness the power of ongoing technological innovations in armor, propulsion, and weaponry. While not as expensive as the subsequent carrier warfare innovation, battlefleet warfare still required a significant financial investment. On the organizational side, while battlefleet warfare represented a continuation of the critical task of navies in the period, winning fleet battles through gunnery, the core competencies required of the average sailor substantially changed, requiring shifts in recruiting, education, and training. Still, the difficulties experienced by most navies in implementing some of the organizational changes necessary to take advantage of battlefleet warfare and the enormous cost of the ships continued to give the British a naval edge well into the twentieth century.

In the years between the launch of the *Dreadnought* and World War I, the British Navy also came close to fully instituting a second, even more wide-ranging naval innovation with the development of the flotilla system. The flotilla system, designed by Fisher to largely replace the battleship with a combination of the destroyer, torpedo boat, and submarine, along with the battle cruiser, is a textbook case of a disruptive innovation. While the more publicized battlefleet segment of the Fisher Revolution was successfully implemented, the more transformational flotilla segment—never fully explained at the time—experienced setbacks as key organizational actors blocked implementation due to uncertainty about effectiveness and bureaucratic politics. Ironically, the last bureaucratic obstacles to the flotilla system were finally overcome in July 1914—although too late to implement it before the outbreak of World War I a month later.

BRITISH NAVAL POLICY IN THE LATE NINETEENTH AND EARLY TWENTIETH CENTURIES

The story of British naval innovation in the late nineteenth and early twentieth centuries is one of transformation from a navy that originally tried to freeze the status quo in the post-Nelson era to a navy that led the way into the industrial age. Scholars have traditionally viewed this change as incremental—a series of technological developments based in phases of the Industrial Revolution leading to changes in the construction, propulsion, armament, and firepower of naval vessels (Herwig 2001; Kennedy 1976). Changes in strategy are seen as the inevitable result of a changing political landscape, like the Anglo-German arms race, and technological developments. Rather than viewing the late nineteenth century as one of straightforward technological developments naturally causing changes in the production of naval power, though, it is more accurate to see the period as one of rapid development in naval strategy that combined new technologies with organizational changes to shift the previous global consensus on naval warfare.

From the 1880s on, when the European land powers, especially Germany, began catching up to Britain in industrial production, the strategic landscape facing the British Navy shifted (Sumida 1989, 6). Potential rival naval powers held back by domestic struggles and financial incompetence in the nineteenth century got their political and financial houses in order, allowing them to start raising and borrowing money on the scale necessary for major naval investments. The pace and cost of technological change for naval power also increased as advances in propulsion, armaments, and armor made new warships obsolete at a faster rate.[1]

Besides the increasing costs of building new ships as the old ones became obsolete faster, the costs of the ships themselves also skyrocketed. HMS *Warrior*, the 1860 ironclad, reportedly cost six times more than an eighteenth-century ship of the line, while the 1888 HMS *Nile* cost more than twice as much as the *Warrior* (Sumida 1989, 8). In 1895, according to Jon Sumida's and Nicholas Lambert's reading of British naval appropriations accounts, British naval spending solely on new battleships was approximately £3.372 million and new warship construction as a whole was £6.198 million. By 1904, spending on new battleships had increased to £4.548 million while total new warship construction spending totaled £11.594 million (Lambert 1999, 306).

The Fisher Revolution as a Military Innovation

Recent historical research by Sumida (1979, 1989, 1996) and Lambert (1995a, 1995b, 1999, 2004), among others, has profoundly changed our knowledge of British naval policy in the late nineteenth and early twentieth centuries. These new historians evaluate the way that financial and political conditions influenced British naval strategy, and their critiques of the traditional evaluations of British naval strategy introduce several relevant points. Arthur Marder and those who followed examined British naval strategy in the late nineteenth and early twentieth centuries through the lens of Mahan and the supposed lessons

---

[1] A Statement of Admiralty Policy 12-1905, box 2, paper 5, Crease [Fisher] Papers, 28, Admiralty 2-15-1906, box 2, paper 13, Crease [Fisher] Papers, 20–21. See also Brodie 1941, 213–15.

of Jutland.[2] The empirical evidence that Mahan influenced the British high command is limited, especially in the decade before World War I. Mahan even came out strongly against the showcase British naval weapon of the period, the HMS *Dreadnought*.[3] When discussing his planned reforms for the British Navy, Fisher opined that Mahan was an "extinct volcano" (Fisher to Fiennes 2-8-1912, Fisher 1956, 430).

We now know, however, that the British Navy was planning for more than decisive fleet battles in the decade before World War I. In addition to the HMS *Dreadnought*, Fisher attempted to implement an ambitious transformation of the British Navy away from reliance on the battleship and toward what he called the flotilla, or the use of torpedo boats and submarines to protect the British homeland, which would enable rapid power projection abroad by a new class of naval warship, the battle cruiser (Lambert 1995a, 648; Sumida 2006, 94–96). Fisher believed that the emphasis on the battleship had created costly military inefficiencies that led to larger numbers of obsolete ships in service. His support for ships like the *Dreadnought* was a means to his end of a much deeper shift in British naval strategy. Yet Fisher's attempt to implement his flotilla defense strategy was frustrated; the navy that the British took to war in 1914 resembled more closely the battleship navy favored by Fisher's detractors than the one that Fisher had envisioned. While naval leaders did "get" the flotilla innovation by mid-1914, World War I prevented its full implementation (Lambert 1995b, 625–26; Lambert 1999, 303; Lambert 2004, 272–73, 286–87).

Fisher was adamant about the adoption of his entire 1904 fleet reform proposal, submitted when he assumed the role of first sea lord, the military head of the British Navy, writing "The Scheme! The Whole Scheme!! And Nothing But the Scheme!!!" in response to those favoring implementation of only part of the plan (Naval Necessities, vol. 1, 1904, Fisher 1960, 20). The multifaceted nature of Fisher's plan makes it appropriate to separate British naval innovations during the Fisher period into two categories: innovations related to the battlefleet, including the *Dreadnought* and battle cruiser, and the flotilla innovations, those related to the significance of destroyers, torpedo boats, and submarines in the defensive and offensive plans of the British Navy. Some of the organizational shifts—such as fleet redeployments and personnel changes, including training and staffing—fall into both categories since they helped enable both sets of innovations.

Lambert's conclusion (1999, 303) that the flotilla innovation was not "apparent to all" highlights both the incompleteness with which Fisher's flotilla scheme was implemented and the lack of international perception. Given the

---

[2] While British Navy officers apparently read Mahan, his views only served to reinforce existing policy paths (Crowl 1986, 473; Hattendorf and Jordan 1989, 350–51).

[3] See memo by William Sims, cited in Fisher 1960, 343–77.

mismatch between the intentions of Fisher and some of his successors with regard to British naval strategy and the signals that the British Navy actually sent around the world, it is necessary to focus on the policies actually implemented by the British Navy, not solely the scheme proposed by Fisher. This section will lay out the key facets of what will be referred to as the battlefleet and flotilla innovations, but the subsequent discussion will concentrate almost entirely on the application of the theory to the battlefleet innovation.

The Battlefleet Innovation

Battlefleet warfare was a military innovation because it represented an enormous change in the way that large navies attempted to conduct what they viewed as their primary mission: massing force to destroy enemy fleets with gunfire, thereby enabling control of the seas. The technological components of the innovation—the HMS *Dreadnought*, subsequent battleships, and the battle cruiser—have come to symbolize the entire period of naval innovations.

While Fisher was the commander in chief of the Mediterranean fleet, he and W. H. Gard, the chief constructor at the Malta dockyards, began working out schemes for the HMS *Untakeable*, a precursor design to the *Dreadnought*, and the HMS *Unapproachable*, a precursor design to the Invincible-class battle cruisers (Fisher 1920, 127, 128; Parkes 1970, 468).

Compromises with the Committee on Designs in winter 1904–5 led to the recommendation for construction of the *Dreadnought* in late February 1905. By this time, after his appointment as first sea lord, Fisher was trying to bring about a much larger shift in British naval doctrine, defined by the battle cruiser (Lambert 2004, 272–73). The *Dreadnought*, though it became his legacy, was in many ways the compromise that allowed him to also build battle cruisers. The ship was laid down on October 2, 1905, launched for the first time on February 10, 1906, and completed by December 1906 (Marder 1964, 532–35).[4]

While one could argue that the *Dreadnought* was the logical conclusion of naval developments over the preceding few decades, this observation is far clearer in hindsight.[5] In its day, the *Dreadnought* provoked an extraordinary backlash related to both its cost and effectiveness, with Fisher saying in retrospect that there was "unanimous naval feeling against the *Dreadnought* when it first appeared" (Marder 1964, 536). Two things separated the *Dreadnought*, and even more so the Invincible-class battle cruisers, from previous British

---

[4] Dispatches by the British naval attaché to Japan, Captain William C. Pakenham, described how Togo's destruction of the Russian fleet favored large uniform-caliber armaments. Given Fisher's interest in the *Dreadnought* prior to the Russo-Japanese War, however, it seems that Pakenham's dispatches reinforced, rather than determined, the *Dreadnought* decision (Committee on Designs 1-3-1905, Fisher 1960, 217–18; Marder 1964, 530; Pakenham to Selborne 11-28-1905, Selborne Papers, box 92, folios 61–66).

[5] For an example of the "logical evolution" argument, see Gardiner, Gray, and Budzbon 1985, 21.

ships. First, instead of placing both large- and small-caliber guns on the deck, Fisher placed only large-caliber guns on the deck. Future battleships would also surround themselves with support ships like destroyers and cruisers to handle smaller threats. Second, rather than splitting the navy into "fast" ships and battleships with heavy armor and guns but little speed, Fisher pushed to maximize the speed of the HMS *Dreadnought*.[6] New turbine engines allowed for an unprecedented cruising speed of 21 knots. It meant the *Dreadnought*, the most powerful naval weapon in the world, was also one of the fastest, able to outrun all but a few cruisers and torpedo boats, to say nothing of other battleships.

The principles underlying the *Dreadnought* also applied to the battle cruiser, especially the large uniform-caliber armament scheme and the emphasis on speed, which Fisher called "one great essential" of modern naval warfare (Naval Necessities, vol. 1, 1904, Fisher 1960, 45). The Invincible-class ships boasted a top speed of almost 25.5 knots. The battle cruiser was the tip of the new imperial defense spear. It was smaller and less heavily armed than the battleship, leading to speed increases that Fisher stated would allow for rapid global deployment (Naval Necessities, vol. 1, 1904, Fisher 1960, 40–45). Given Britain's increasing dependency on foreign imports, especially food, fast battle cruisers would enable Britain to both protect its merchant ships and threaten the commerce of potential competitors (Lambert 1995b, 598).

The set of innovations usually symbolized by the *Dreadnought* encompassed much more than the famous ship. The development of wireless telegraphy, first demonstrated in warfare by the Japanese in the Russo-Japanese War in 1904–5, created new opportunities and challenges for coordination at sea that further influenced naval planning. While the technology was not advanced enough to allow for the real-time direction of battle from land, it did improve communications within fleets at sea. This allowed for faster and more reliable coordination of movements. For example, direct communication between ships and the shore within a five-hundred-mile triangle of British naval bases at Malta, Gibraltar, and Cleethorpes (on the British coast) became possible by 1906 (Lambert 2005, 379). The wireless telegraph also helped scout ships to report on enemy positions, changing strategies about tracking and planning for how to intercept and destroy the enemy fleet.

The Fisher period involved a series of organizational changes as well. In 1902, First Lord Selborne, the civilian head of the British Navy, began implementing a series of changes in the way that the British Navy recruited, educated, and trained its sailors, known as the "Selborne Scheme." These changes shifted everything from the way that future officers and enlisted personnel were recruited into military to the relationship between engineers and officers to the

---

[6] See Admiralty Policy in Battleship Design 1906, Speed of Battleships 11-2-1906 (Fisher 1960, 314, 341–43).

way the military trained to fight. First, the practice of accepting volunteers at young ages—twelve or thirteen years old—was reinitiated. Fisher believed that the increasing complexity of modern fleet operations required more training time. He also attempted to widen the naval recruiting net to draw in more officers from the rising middle class, rather than just aristocrats who thought a naval career was a good way to see the world and avoid work. As Fisher once said, "We are drawing our Nelsons from too narrow a class" (Marder 1961, 30–31).

Second, in a nod to U.S. naval training, Selborne and Fisher revised the distinction between engineers and officers. Previously, the two groups attended separate schools, came from different social classes, and progressed on different promotion tracks. As the role of the engineer increased in importance and a deep engineering background became necessary for effective decision making by officers, the old distinctions no longer made sense. The training schools were merged, and promotion paths were consolidated.[7]

Third, Fisher created a nucleus crew system that increased readiness by giving reserve ships larger complements even during periods of inactivity and preplanning the distribution of personnel in the case of a large-scale mobilization.[8] Fourth, in line with growing pressure from private innovators like Arthur Pollen and high-ranking naval officers like gunnery expert Percy Scott, Fisher designed a more realistic training plan, including more long-distance firing (Herwig 2001, 121; Sumida 1989; Sumida 1996).[9] Fisher in general forced a higher degree of realistic exercises on the fleet. Instead of "training" by swabbing the deck and climbing the rigging, Fisher emphasized training by practicing fighting the way that the naval leadership thought it would actually happen on the battlefield. Echoing Nelson's credo that "the battle ground should be the drill ground," Fisher required more battle practice and maneuvering (Marder commentary in Fisher 1956, 24–25). While some admirals felt that he disrespected the "old ways," Fisher actively encouraged what today would be called thinking outside the box. Fisher even sponsored prizes for innovative ideas in naval strategy written by junior officers (Massie 1991).

Finally, in 1904, the British Cabinet and Board of Admiralty approved a major redistribution of British naval forces, withdrawing and then scrapping over one hundred cruisers and gunboats from British naval stations abroad.

[7] While this generated an enormous outcry among many retired naval personnel and even some active admirals, Selborne believed the new scheme was necessary to effectively train the next generation of British officers (New Scheme of Naval Education 4/1905, box 1, paper 11, Ewing Report on Visit to America 12-2-1904, box 1, paper 19, Crease [Fisher] Papers; Selborne, Training Memorandum 12-1902, 161/25, Selborne Papers).

[8] Navy and Dockyards, 11-14-1905, box 2, paper 2, Crease [Fisher] Papers; Naval Necessities, vol. 1, 1904, Fisher 1960, 45–48; Naval Necessities, vol. 2, 1905, Fisher 1964, 20–24; Fisher to Tweedmouth, 5-29-1906, Tweedmouth Papers, 254/95.

[9] Scott began designing more realistic, gunnery practice when he took over the Gunnery School on Whale Island in 1903 (Jones and Keogh 1985, 125).

This redeployment has come to symbolize a larger change in British foreign and military policy away from imperial concerns, and toward the rising German threat (Friedberg 1988; Kennedy 1976). In contrast to the belief that the redeployments represented a shift in British focus from imperial affairs to Germany, though, the redeployment was not a change in the goals of the British Navy but rather in the means used to achieve those goals. The redeployment shifted the British Navy to a rapid deployment maneuver strategy designed to more effectively protect British imperial interests than the forward presence strategy while also cutting costs (Lambert 2004, 285).[10] The focus on Germany came later. In particular, the withdrawal cut obsolete ships from the navy list, producing enormous cost savings, and more efficiently used the crews through the nucleus system, which kept a much higher percentage of active British naval ships able to deploy and fight quickly.[11]

The term battlefleet innovation therefore refers to the integration of a series of incremental technological advances with the organizational changes described above. It was a sustaining innovation because it did not disavow the idea of battleship battles or explicitly repudiate the idea of the decisive engagement. Much more important than the technological advances was the way that the British Navy combined the various technologies together and reformed the naval bureaucracy to best take advantage of them. In combination with organizational changes that overhauled the staffing of ships and the way they trained, the new battlefleet generated a dramatic improvement in the capacity of the British Navy to produce military power both regionally and globally.

A report in 1913 by Rear Admiral Mark Kerr, the British naval attaché to Greece, shows the significance of considering the battlefleet warfare innovation as it was perceived in the international community at the time. Part of Kerr's job involved consulting with the Greek Navy about the best way to ensure effective control of the Aegean. While he attempted to push the development of a flotilla strategy based on the submarine and torpedo, knowing that it represented cutting-edge thinking in the British Navy, his Greek counterparts rejected his proposal. Instead, they emphasized the importance of having dreadnought-quality battleships as a marker of naval power.[12]

The Flotilla Innovation

Fisher also advocated a system of flotilla defense, whereby destroyers, torpedo boats, and the submarine would replace the battleship as the primary

---

[10] For an application of this argument to deployments in China, specifically, see Naval Necessities, vol. 2, 1905, box 1, paper 7, Crease (Fisher) Papers, 153–59.

[11] Naval Necessities, vol. 1, 1904, Fisher 1960, 45–48; Naval Necessities, vol. 2, 1905, Fisher 1964, 20–24; Fisher to Tweedmouth, 5-29-1906, 254/95, Tweedmouth Papers. For more on the redeployment in 1904, see Horowitz 2008.

[12] Documents authored by Kerr, cited in Lambert 1999, 303.

mechanism for defending the British homeland from invasion by sea. Fisher thought the increasing range, accuracy, and power of torpedoes, along with developments in the submarine, meant that the age of the battleship, or even the large-scale capital ship, was ending.[13] This was especially true in the narrow English Channel, where effective flotilla defenses would free British battleships for more effective duty elsewhere, either engaging enemy fleets, protecting British commerce, or protecting other British interests abroad.

The flotilla represented an entirely different way of thinking about naval warfare—closer to the Jeune École than the popular interpretation of Mahan. Swarms of torpedo boats, destroyers, and cruisers, supported by some larger ships, could attack enemy vessels at sea or conduct coastal defense operations. Even with a full organizational commitment, full funding would have required an even more massive redistribution of fleet resources and ships. This was a fundamentally disruptive innovation. While the battlefleet innovation overturned old core competencies like the melee, climbing the mast, and targeting the guns by eyeballing the target, it was consistent with the critical task of using the guns of large capital ships as the primary element of naval power to fight and win battles. Flotilla defense reversed the emphasis on fleet battles at home, making it inconsistent with the preexisting critical task of the British Navy (Lambert 2004, 278–79; Lambert 2005, 376).

It is, of course, impossible to predict how history might have changed if it had been fully implemented and noticed before World War I; it could have failed, giving the Germans control of the seas during World War I, or succeeded, making the British both less vulnerable to attack by the German fleet and more capable of handling the German U-boat threat. But since flotilla warfare never "debuted," the rest of this chapter will examine battlefleet warfare.

The Debut of Battlefleet Warfare

The natural place to code the "start date" for the battlefleet innovation is the point at which other navies around the world recognized that the British had begun doing something different. This occurred during the widespread leaks about the HMS *Dreadnought* that spread in the British media and beyond in early 1906. In an era notorious for the open flow of information, the British Admiralty kept a tight lid on details concerning the *Dreadnought*. Production

---

[13] In one letter Fisher stated that submarines would magnify Britain's naval power by seven times (Fisher to White, 3-12-1904, Fisher 1952, 305). In another letter he wrote, "It's astounding to me, *perfectly astounding*, how the very best amongst us absolutely fail to realize the vast impending revolution in naval warfare and strategy that the submarine will accomplish." Fisher also said submarines would cause an "immense impending revolution" as "offensive weapons of war" (Fisher to May, 4-20-1904, Fisher 1952, 308–9).

itself was kept a secret for as long as possible, with only a few newspaper leaks providing any information of British construction plans until the first launch of the ship in February 1906 (Marder 1964, 535–36). Even after news of the ship became public, the Admiralty shielded details about its actual capabilities from the public, with the idea that if the true capabilities and reasoning behind the production of the *Dreadnought* remained secret, it would help increase the confusion abroad regarding British naval planning, increasing Great Britain's relative naval edge (Bacon 1929, 254; Marder commentary in Fisher 1956, 31; Marder 1964, 538–39). The eventual spread of information about the *Dreadnought* in early 1906 spurred large-scale reactions by other global navies, delaying the production plans of navies around the world for over a year (Bacon 1929, 247). So while realization of the battlefleet warfare innovation in Great Britain began in 1902 with the first stages of the Selborne Scheme and continued after Fisher became first sea lord in 1904, the debut point did not occur until knowledge of the *Dreadnought* spread. Unlike many other cases, where the critical debut for an innovation was its use in a war, the critical introduction for battlefleet warfare was the release of information about the ship even before its use in a military context.

Predicting the Spread and Impact of Battlefleet Warfare

Required Financial Intensity

Adopting the battlefleet innovation required a large degree of financial capacity on the part of potential adopters. In Great Britain, the relative cost per unit of the elements of British naval power skyrocketed in the late nineteenth and early twentieth centuries (Lambert 1999). The HMS *Warrior* and *Black Prince*, twin ironclads laid down in 1859, cost about £378,000 each. Costs had more than doubled by 1874, when the *Inflexible* was laid down at a cost of £812,485. While costs briefly declined with the production of ships like the *Renown*, laid down in 1893 for about £750,000, they climbed once again as warship size and armaments continued to increase along with the demands for more speed. The pre-*Dreadnought* ships of the King Edward VII class, laid down between 1902 and 1904, cost about £1,350,000, while the last two pre-*Dreadnought* ships, the *Lord Nelson* and *Agamemnon*, laid down in mid-1905, cost about £1,550,000.[14] The *Dreadnought* itself, laid down on October 2, 1905, cost £1,783,883, an increase of almost five times the cost of the *Warrior* and over 30 percent in comparison with the King Edward VII class. The battle cruiser was also not a cheap

---

[14] The financial statistics are from Oscar Parkes, whose numbers vary slightly from Fisher's own (Admiralty Cost in Battleship Design, 1906, Fisher 1960, 341; Parkes 1970, 425, 451, 477). To give a sense of the relative cost, converted from 1906 pounds into 2005 dollars, the *Dreadnought* cost about $185 million (Officer and Williamson 2006).

ship; the three Invincible-class ships laid down after the *Dreadnought* in early
1906 cost over £1,600,000 each (Parkes 1970, 492). Add in the cost of wireless
communications technology, training for engineers, and other technical specialists newly required for industrial navies in comparison with their sailing
predecessors, and the cost per unit required to implement the innovation was
medium.

The underlying technologies of the innovation are a mix between those
designed for commercial and military purposes. Specifically, innovations like
wireless communications and turbine engines, though heavily supported and
subsidized by national militaries, at least in the British case, also presented
clear commercial opportunities. On the other hand, armor and armaments
were entirely military subcomponents.[15] Therefore, in this case—the cost of
each unit of power—the dreadnought- or Invincible-quality ship, combined
with the nature of the underlying technologies, dictates a financial intensity
rated as medium.[16]

Required Organizational Capital

The organizational capital required to implement battlefleet warfare, for potential emulators, is higher than some have assumed. Because traditional histo-
ries have focused on the technology of the *Dreadnought* itself or the influence
of great leaders like Fisher, Alfred von Tirpitz, and Togo Heihachiro, organizational development has been less scrutinized. The organizational and planning reforms begun when Fisher overhauled the personnel system as second
naval lord in 1902 and continuing when he took over as first sea lord in late
1904 represented profound changes not only in how the British Navy recruited
sailors but also in what they were trained to do.

Battlefleet warfare involved training to fight at much longer range and potentially without decisive fleet battles. Given advances in the torpedo, propulsion,
and armaments, the optimal range of engagement required by the innovation
was several thousand yards longer than the previous norm of just a few thousand (three thousand was the norm in 1903). Fisher had already experimented
with longer-range battle practice as commander in chief in the Mediterranean
in 1899 (Sumida 1996).[17] Second, the development of wireless telegraphy meant

---

[15] For a broader discussion of developments in naval technology during the period, see Brodie
1941, 9–13.

[16] This case breaks out of the ideal types for coding financial intensity described in chapter 2, as
there are some indicators pointing in each direction. While there were high unit costs for adoption, the relative increase was not extremely significant. Additionally, although much of the underlying technology was military, some was civilian.

[17] The French and Italian navies also conducted experiments during that period (Parkes 1970,
466).

military organizations had to develop the capacity to handle a larger quantity of data and simultaneously coordinate the actions of a larger number of ships, leading to new demands on intelligence organizations along with some shifts in the relationship between ships at sea and the naval bureaucracy at home (Lambert 2004, 283–84). Finally, as described above, taking advantage of British naval innovations in the period meant dealing with the complicated relationship between commanding officers and engineers.

In the years prior to the debut of battlefleet warfare, Great Britain demonstrated a medium level of organizational capital. The late nineteenth-century British Navy viewed its critical task as controlling the sea to protect the British homeland and imperial lines of communications to its colonies. The ability to win decisive fleet battles represented the means of achieving sea control throughout the late nineteenth century (Friedberg 1988). Defining sea control as the ability to win decisive battles did narrow the task focus of the navy. Given that battlefleet warfare fell within the rubric of ways to win decisive fleet battles, however, the focus of the British Navy did not substantially hinder its ability to adopt the innovation.

The British Navy in the late nineteenth century did have an "old" organizational age, signaling a relatively low level of organizational capital. As the dominant global naval power since the early nineteenth century, having arguably not experienced a defeat since the Battle of the Chesapeake in 1781 during the American Revolution, the late nineteenth-century British Navy had little incentive to change in fundamental ways (Lambert 2004, 275–76). Advocates of the Selborne Scheme claimed that previous "tinkering" had been "futile attempts to adapt a system wrong in principle and incapable of assimilating modern developments to present-day requirements" (Shore 1906, box 2, paper 8, Crease [Fisher] Papers).

Lambert and others have shown that the British Navy was not inherently anti-innovation in the late nineteenth century but instead was devoting resources to experimentation and research (Brooks 2005; Edgerton 1996; Lambert 1999). But as Lambert (2001, ix–xx) contends, the British approach to new weapons such as the submarine was to carefully and quickly follow developments by first movers like the French, rather than lead the way themselves. In areas like propulsion and armaments, the British Navy closely tracked the work of other navies, and then attempted and nearly always succeeded in implementing the technologies themselves (Fyfe 1902, 281). Still, while the British did invest heavily in research and attempted to maintain the ability to stay on the cutting edge of naval developments, they generally did not heavily invest in pure experimentation (Massie 1991; Sumida 1989, 1996). As the leading naval power, they viewed it as unnecessary.

Using the first mover, Great Britain, as a marker for the general level of organizational capital required to adopt the innovation, the data suggest a medium level required to implement the innovation.

Battlefleet Warfare Diffusion Predictions

Given the medium levels of financial intensity and organizational capital required to adopt battlefleet warfare, the general system-level prediction is a stable but not overwhelmingly high rate of diffusion to a sizable number of states, with many states pursuing dual strategies of attempting adoption in one form or another along with countering or an alliance strategy.

Battlefleet Warfare Impact Predictions

As an innovation requiring a medium level of both financial intensity and organizational capital to implement, the innovation should mostly benefit major powers, but also give smaller powers an opportunity to rapidly build naval capabilities. The sufficient distinctions between the all-big-gun battleship and those that came before it should give the first mover an advantage—but a short one, since the financial requirements are not so overwhelming that other states cannot compensate and build the technological components of the innovation. The battlefleet warfare case does not fit as neatly as some of the other cases because of the medium required level of financial intensity and organizational capital to adopt the innovation.

The relative costs of building a battleship should not decline over time given the pace of technological change during the period. Therefore, waiting to adopt the innovation should not necessarily give later adopters any technological advantages. Since the level of organizational capital required to adopt the innovation is also medium, though, later adopters might benefit from learning how earlier ones attempted to implement and actually use the innovation.

The theory also makes a number of predictions with regard to the relationship between innovation diffusion, power, and war that should apply in the battlefleet warfare case.

> **Major Power Dropout Hypothesis:** Major powers that fail to adopt battlefleet warfare will gradually cease to be major powers in the area of the innovation.

> **Military Interaction Hypothesis:** States that adopt battlefleet warfare, even holding other things constant, should be involved in more militarized disputes and more intense ones with a wider geographic set of states than nonadopters.

The Diffusion of Battlefleet Warfare

The context surrounding the emergence of battlefleet warfare during the pre–World War I period and the aftermath makes quantitative statistical tests difficult. First, the innovation was only relevant for a short period of time—from its initial debut in 1905–6 through the beginning of World War II.[18] Its diffusion was also more or less completed by the end of World War I, leaving a short period of time from which to draw observations for statistical analyses. Second, most of the actual fighting involving the battlefleet occurred during World War I. Traditional data sets like the MIDs and COW ones do not adequately depict the complex interactions endemic to a conflict such as World War I. Fortunately, it is possible to inspect the universe of cases.

Preinnovation Naval Powers

While Great Britain was the preeminent global naval power of the late nineteenth and early twentieth centuries, a set of several other countries also possessed large navies and represented, in combination, a potential challenger to Great Britain. This drove the British Navy to construct the "two-power" standard in the 1880s, which stated that the British Navy would maintain forces capable of defeating a simultaneous attack by the next two strongest powers. British naval planning during the period focused on France and Russia as the most likely enemies of Great Britain, even after the signing of the entente cordiale in April 1904.[19] Table 5.1, taken from a document for the British Parliament prepared by First Lord Selborne in late 1903, shows the distribution of relative naval power in the year before the Russo-Japanese War.

Conway's *All the World's Fighting Ships*, the most authoritative source on global navies for that period, defines the major naval powers of the 1906–21 period as countries with a theoretical capability for blue-water naval operations: Great Britain, the United States, Germany, France, Italy, Japan, Russia, and Austria-Hungary. Coastal naval powers, or those naval powers with some fleet capabilities but mostly organized for the purposes of defending their coastlines, were Norway, Denmark, Sweden, the Netherlands, Portugal, Spain, Greece, Turkey, China, Argentina, Brazil, Chile, and Peru. While many other countries—even Zanzibar—had nominal navies, they had minor investments in naval capabilities (Gardiner, Gray, and Budzbon 1985, forward). The major and coastal navies are the universe relevant for this chapter.

Diffusion Results

When word leaked of the HMS *Dreadnought* in early 1906, the construction programs of virtually every other global naval power paused in response to Great Britain's actions. Countries altered the construction plans for already-planned

---

[18] For more on the debut of the "next" innovation, see chapter 3.

[19] For the text of the entente cordiale, see United Kingdom and France 1904.

TABLE 5.1
Estimated Distribution of Naval Power on January 1, 1904

Country	Battleships (1st and 2nd class)			Modern armored cruisers or corsairs		
	Built	Building	Sanctioned but not yet laid down	Built	Building	Sanctioned but not yet laid down
France	29	6	—	13	5	3
Russia	21	8	—	5	3	—
Total—France and Russia	50	14	—	18	8	3
Great Britain	51	12	—	15	17	—
Japan	7	—	3	6	—	3
United States	11	11	3	4	11	—
Germany	19	5	22	2	2	11

Country	Destroyers			Torpedo boats of 117 feet and above		
	Built	Building	Sanctioned but not yet laid down	Built	Building	Sanctioned but not yet laid down
France	24	9	6	181	20	126
Russia	49	11	—	85	—	—
Total—France and Russia	73	20	6	266	20	126
Great Britain	116	30	—	85	4	—
Japan	19	1	—	57	12	—
United States	18	2	—	22	4	—
Germany	31	6	78	88	—	—

Country	Submarines		
	Built	Building	Sanctioned but not yet laid down
France	16	24	34
Russia	—	4	—
Total—France and Russia	16	28	34
Great Britain	6	12	1
Japan	—	—	—
United States	8	—	—
Germany	—	1	1

Source: Selborne 12-7-1903, Selborne Papers 161/XXVIII.

warships and began looking closely at the new British one. For example, the size and depth of dreadnought-style ships not only required a complete change in German construction plans but also required the widening and deepening of the Kiel Canal so that large battleships could pass through into the North

Sea.[20] After alterations began, Fisher (1956, 424) viewed them as so important that he timed his prediction for the date of the Battle of Armageddon to coincide with German completion of the project. In fact, Germany completed the canal enlargement project in late June 1914, barely a month before the war started. The introduction of the *Dreadnought* forced a range of naval policy changes.

- France: Did not lay down a dreadnought-quality battleship until 1910, launching the mixed-caliber Danton class in between 1906 and 1910.
- Germany: Forced into multimillion dollar redesign of the Kiel Canal. First dreadnoughts, the Nassau class, were not laid down until June 1907.
- Japan: Forced to convert Satsuma-class battleships to mixed caliber in spring 1905 due to cost considerations. Kawachi-class battleships incorporating *Dreadnought* insights were not authorized until 1909 and were not completed until 1912–13.
- United States: While the *Michigan* and *South Carolina* were envisioned in May 1905 with uniform large-caliber armament, internal disagreements delayed the awarding of contracts until July 21, 1906, and the ships were not deployed until 1909. Delaware-class battleships designed post-*Dreadnought* were not authorized until 1907 and were not deployed until 1910.
- Austria-Hungary: Commander in Chief Rudolf Montecuccoli secretly arranged for the construction of dreadnought-quality battleships in late 1909, and the *Viribus Unitis* came into service in 1912.
- Italy: The first Italian dreadnought, the *Dante Alighieri*, was not laid down until early 1909 and was actually finished after the *Viribus Unitis*, despite being authorized and sailed earlier.
- Russia: No dreadnoughts constructed until the Gangut-class ships, which were authorized by Czar Nicholas II in late 1908, laid down in 1909, and completed in 1914.

Figure 5.1 shows the spread of British naval practices following the demonstration point of the innovation: widespread knowledge concerning the basic capabilities of the HMS *Dreadnought* in early 1906.

The spread of the technological components of the innovation, like dreadnought-style battleships, occurred much more quickly and completely than the spread of the complete innovation. Given the medium levels of financial intensity and organizational capital required to adopt, the expectation of a fairly consistent spread of the innovation, with the technology spreading more quickly, is borne out by the evidence. Since the technological side of the innovation

---

[20] The Kiel Canal connected Germany's Baltic bases to the North Sea, which allowed German warships to enter the North Sea without moving through international waters. It was originally completed in 1895.

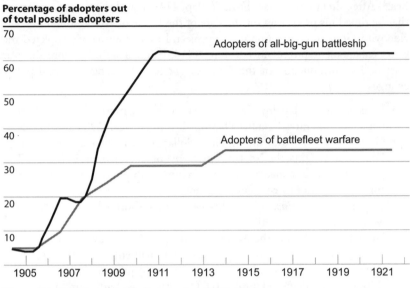

**Percentage of adopters out
of total possible adopters**

Figure 5.1. Diffusion of battlefleet warfare, 1905–21.

should spread relatively quickly to those with the financial capabilities, it is not surprising that the advantage for the first mover, Great Britain, is shorter in this case than for harder-to-adopt technologies. Yet on the organizational side, the delay between the British innovation and widespread adoption is longer.

THE IMPACT OF BATTLEFLEET WARFARE

On reputation alone, few military innovations have had as large an impact on the security environment as the HMS *Dreadnought*. From the beginning, the British Admiralty believed that "the leap forward would prove of great value" (Admiralty 10-1906, box 2, paper 22, Crease [Fisher] Papers, 21). Led by Fisher, the Admiralty attempted to maximize the first-mover advantage gained from building the *Dreadnought* by limiting the information released about the ship—even though it made its task of dealing with critics in the British media more difficult. As Fisher wrote,

> It is obviously undesirable that the Board of Admiralty should lay bare in a public statement the motives which underlay their policy of ship con-struction. To do so would give foreign rivals the whole of the benefit of the experience which has been too laboriously acquired at our Admiralty and dockyards, and in our fleets while a complete exposition could hardly be published without giving dire offence to certain of our potential ene-mies. (Admiralty Policy in Battleship Design, 1906, Fisher 1960, 303)

The Admiralty purposefully shielded the results of battle trials that showed faster speeds and more accurate targeting at longer distances, believing that secrecy would delay the efforts of foreign navies to mimic the *Dreadnought*, helping sustain and widen Britain's relative naval advantage (Bacon 1929, 254; Marder commentary in Fisher 1956, 31; Marder 1964, 538–39).[21]

The British did gain important short-term advantages from their first-mover status. Despite the lack of "proof of concept," or actual success in battle, the *Dreadnought* almost immediately became the gold standard of naval power, so much so that the entire generation of ships is separated into pre-dreadnought and dreadnought ships, with the name dreadnought coming to mean not just the ship itself but also the package of capabilities it contained (Gardiner, Gray, and Budzbon 1985).

The introduction of the *Dreadnought* delayed the naval construction programs of the other major naval powers to a degree even beyond that anticipated by the Admiralty (Admiralty 10-1906, box 2, paper 22, Crease [Fisher] Papers, 21). Given the rapidly agreed-on standard whereby all ships prior to the *Dreadnought* became inadequate, Great Britain jumped out to a decisive lead in naval power. For example, while France's relative naval fortunes, as explained below, had been in decline throughout the preceding decade, the battlefleet warfare innovation likely made France even more dependent on Britain and more willing to offer concessions. The innovation also gave Britain a short-term negotiating edge, as the Morocco crisis soon demonstrated.

These advantages were short-lived, however. Within a few years the Germans had begun construction of their own dreadnought-quality battleships, and started widening and deepening the Kiel Canal. The other major naval powers also began building dreadnought-quality battleships, and even Italy authorized construction of its own dreadnoughts by 1909. The declining numbers of ships in the fleet, due to the rising costs, arguably made it easier for navies like the Germans to gain on the British Navy, though the British Navy did manage to stay ahead. So while Great Britain gained a large short-term relative power advantage, the adoption requirements shortened the first-mover advantage experienced by the British Navy.

The implementation of Fisher's organizational changes, especially the nucleus crew system and changes in battle practice to make it more realistic, also increased the fighting efficiency of the British Navy. For instance, the fleet redeployment of 1904 helped Fisher eliminate costly and outdated vessels and deployments, and reinvest the money in faster battle cruisers and battleships that would more effectively allow for sea control and imperial

---

[21] The success in delaying the naval construction programs of other navies bolstered Fisher's belief in the idea of "plunging," or attempting to introduce large-scale changes to the British naval construction program at critical junctures to induce confusion and further disrupt construction by the other naval powers (Lambert 2004, 280).

defense.[22] The British concluded that the care and maintenance of aging ships imposed a larger cost on military readiness than the potential reputational benefit of permanent overseas force deployments, and acted accordingly.

At the system level, the impact of battlefleet warfare was mostly restricted to the interaction of major powers. Put crudely, the rich got richer and the poor got poorer, generating higher levels of asymmetries between major power adopters and minor naval powers. The battlefleet innovation shifted the speed of interactions along with expectations of the timing and outcome of battles, which had spillovers into broader strategic planning. The naval arms buildup between Great Britain and Imperial Germany that occurred during the early period of the battlefleet innovation is perhaps the canonical example of an arms race with all the attendant dangers, including capabilities-based planning that presumes the worst about adversaries and leads to an inevitable ratcheting up of tensions (Huntington 1957).

Additionally, with more relative naval power riding on the hulls of individual ships as the cost per unit increased, the military cost of adversaries gaining complete knowledge about the ship increased as well. The growing complexity of naval operations also led to a decrease in naval interoperability. In the age of sail, battles like Navarino in 1827 showed that while cooperation between the English, French, and Russian navies was not easy, it also did not require extensive preplanning. The ships all moved in the battle space more or less in sight of the others, and thus the potential variation in tactics and physical differences between ships was relatively small. In the machine age, all of those factors changed; large variances in strategy, ship capacity, and the size of the battle space made it harder for navies to cooperate without more extensive preplanning. Coordination difficulties between the entente powers in World War I, for example, had a substantial impact on strategy in the Mediterranean. The Mediterranean had to be split into zones of control patrolled by the British, French, and Italians (after their entry into the war). The entente powers lacked the ability, for the most part, to conduct joint operations or even effectively communicate (Halpern 1994; Pack 1971, 162; Tritten and Donolo 1995, 58–59).

How Countries Responded to Battlefleet Warfare

Country-Specific Predictions

The section on alternative strategic choices above predicts that, given the medium levels of financial intensity and organizational capital required to adopt the innovation, states should engage in a wide variety of external and internal

---

[22] Navy and Dockyards 11-14-1905, 6–7, 14–15, 18, box 2, paper 2, A Statement of Admiralty Policy 12-1905, 28–30, box 2, paper 5, Remarks on Memoranda 1-1907, box 3, paper 4, Crease [Fisher] Papers. See also Friedberg 1988; Lambert 1999, 104, 101–15; Lambert 2004, 284–85.

response strategies to battlefleet warfare. Table 5.2 lays out country-specific predictions and results regarding the diffusion of British naval practices. The critical task, organizational age, and experimentation columns reflect the measurement schemes for organizational capital described in chapter 2.

Specific Cases

While the system-level results provide support for adoption-capacity theory, a more fine-grained analysis of four key states will help flesh out the specific effects of battlefleet warfare. The pre–World War I German Navy has been analyzed at length, and the strategic imperatives for the German Navy outlined by Admiral Tirpitz make the German decision overdetermined historically, in contrast to some of the other cases (Herwig 1997). Yet it is important to lay out how this theory can explain the German case as well. France did not immediately or completely adopt the innovation, but it had a clear strategic interest given its understanding of its global role. Japan possessed the organizational capacity for adoption and was advanced enough that it might have created the innovation on its own given more time. Finally, while most people forget that the Austro-Hungarian Empire had a navy at all, the Austro-Hungarian Navy did construct several dreadnought-quality battleships before the First World War. The cases below show that adoption-capacity theory is useful for understanding the state level as well as the system level.

GERMANY

After eighteen years of rule by army generals, the German Navy reassumed control over its own budget and trajectory in 1890, beginning a significant naval modernization program just a few years later (Massie 1991, 162). As scholars from a variety of theoretical backgrounds have pointed out, several factors—including Germany's natural growth in material power and interests, the personality of Wilhelm II, and domestic politics—accounted for growing interest in building a significant navy (Herwig 1987; Hobson 2002). Michelle Murray's recent research (2009), for example, shows how Germany's search for an identity and status within the international system helped drive its naval expansion. The philosophy of the German Navy in the pre–World War I period was infamously articulated by Tirpitz, the head of German's Imperial Naval Office in the prewar period, as the creation of a "risk fleet" that could challenge British naval supremacy in the North Sea if a war occurred (Massie 1991, 181). The idea was to build up a fleet strong enough that the British, with naval commitments around the world, could not risk war with Germany since the costs of defeating the German Navy would place the remainder of the British Navy at risk from France and Russia. The German Navy Bill of 1900, which drew the attention of the British, laid out clearly Germany's intention to build

TABLE 5.2
Country-Specific Battlefleet Warfare Results

	Financial capacity	Critical task	Organizational age	Experimentation	Relationship to first mover at time of demonstration (Great Britain in 1906)	Predicted response to innovation	Accurate prediction
Austria-Hungary	Low	Adriatic Sea superiority; commerce protection in Mediterranean	Negligible—small organization in general	Not in new areas of core competency except the torpedo, but that had been outsourced years before	Cordial	Technological adoption after delay; limited organizational adoption; balancing against first mover; development of counters.	Almost: Theory predicts delayed adoption, but not the lack of serious investment in counters to the innovation
United States	High	Western hemisphere, Hawaii, and Philippines protection	Since 1860s	Yes	No formal alliance but increasing ties	Bandwagoning with first mover; adoption of technology and organizational constraints in short-to-medium term	Yes
Japan	Medium-high	Fleet superiority in the Pacific; coastal defense	Post-Perry visits of 1853–54	Not extensive actual experimentation, but war college system led to war-gaming future naval contingencies	Treaty relationship	Bandwagoning with first mover; adoption of technology and organizational components in short-to-medium term	Partially

Germany	Medium-high	Tirpitz risk theory—fleet engagement	Post-German unification: 1871	Not extensive	Cordial	Technological and organizational adoption in short term	Yes
Italy	Medium-low	Adriatic Sea superiority; commerce protection in Mediterranean	Post-Italian unification: 1861	Some experimentation with long-range gunnery	Stable	Technological adoption after delay; limited organizational adoption; balancing against first mover; development of counters	Yes
France	Medium	Coastal defense; commerce protection in the Mediterranean	Just recovering from instability of Jeune École debate	Experimentation tradition in 1870s to 1880s; none after that	Treaty relationship and increasing ties	Some technological adoption after relatively short delay, but little organizational adoption; bandwagoning with first mover	Yes
Russia	Medium	Unclear	Naval formation by Peter the Great in 1696	No	Treaty relationship with ally of first mover, but participation in war against another ally of first mover	Technological adoption; no organizational adoption	Yes

up its naval power (British Naval Policy Document, 12/24/1903, Selborne Papers, 161/28). While the growth of the German Navy is traced to the 1898 naval bill, a major victory for Tirpitz and Wilhelm II, from an international naval perspective the 1900 bill was crucial since it doubled the number of planned battleships from sixteen to thirty-two (Herwig 1987, 266). Essentially, Germany had already invested heavily by the time the *Dreadnought* debuted. The Royal Navy's director of naval intelligence, Rear Admiral Wilfred Custance, presented data to First Lord Selborne in spring 1902 showing the growth in the German Navy and projections for future growth (Custance to Selborne, 1902, Selborne Papers, 158/41–50).

After World War I, in his memoirs, Tirpitz claimed that he viewed the debut of the HMS *Dreadnought* as an opportunity for Germany since it wiped the naval power slate clean (Herwig 1987, 54).[23] The reality, however, underscored the impact of the *Dreadnought* on international politics. At the time Tirpitz viewed the *Dreadnought* as a significant new challenge to the burgeoning German Navy that threatened the gains Germany had already made. After all, Germany was in the process of building a deeper Kaiser Wilhelm Canal and had to completely retool the project at great expense. In addition, the debut of the *Dreadnought* meant that Germany had, in effect, spent a great deal of scarce resources on ships that would no longer prove relevant.

Nevertheless, as predicted by adoption-capacity theory, while the *Dreadnought* provided a significant short-term first-mover advantage for the British vis-à-vis the German Navy, Germany did have the requisite financial intensity and organizational capital to adopt the innovation and begin to catch up. Germany was a relatively young state, having just been founded in the mid-nineteenth century in the wake of the German wars of unification, concluding with the Franco-Prussian War in 1870–71. As a traditional land power, Germany placed a lower emphasis on the development of its navy after the wars of unification and even through the first decade of Wilhelm II's rule (Massie 1991, 163). But the German Navy had begun more significantly investing in naval assets in the last decade of the 1800s; its development became a budgetary priority. Therefore, the German Navy had a low organizational age and lacked a well-defined critical task, making it well placed to adopt battlefleet warfare—and it probably could have handled something even more organizationally complex as well.

On the financial side, Germany's rising industrial might in the late nineteenth century, showcased by growth in its iron, steel, and energy production,

[23] The German use of submarines, the most effective use of naval power by the Germans during World War I, did not result from careful prewar planning but rather from the realization that the risk fleet had not in fact paid off and was only at risk of being destroyed by the British. Thus, the Germans had to turn to an alternative means to exercise naval influence, leading to submarine warfare of a sort that the French, ironically, had theorized about through the Jeune École, but had not been significantly considered by the major naval powers prior to the war.

gave it the financial capacity to adopt the innovation. Military spending data from the COW Project similarly show an increase in German military expenditures overall from £13.57 million in 1872, the first peace year after the Franco-Prussian War, to £46.16 million in 1905, the last fiscal year prior to the *Dreadnought* (COW military expenditure data are presented in thousands of current-year British pounds for years prior to 1914, Correlates of War 2 Project 2006). Given this overall jump in military expenditures, it is not surprising that Germany's naval budget also increased from 133 million German marks in 1899 to approximately 233 million in the 1905–6 fiscal year (Herwig 1987, 278). Thus, Germany possessed both the organizational and financial capacity necessary to adopt battlefleet warfare, meaning the first-mover advantage gained by the British, though real, did not last. In some ways the German case is overdetermined. Variables from several theories, including those focused exclusively on material resources or domestic politics, predict the development of a larger German Navy in the pre–World War I period. Yet adoption-capacity theory does predict German behavior as well, and given the consistency with the other cases, it suggests the plausibility of the explanation in the German case.

FRANCE

From the mid-nineteenth century through the early twentieth century, France declined from the second-rated global naval power to the fifth (measured by capital ship counts), despite not having a substantial change in its national interests or core financial situation (Gardiner, Gray, and Budzbon 1985, 190). The appointment of Charles Pelletan in 1902 as the minister of marine reignited a debate between the Jeune École, a segment of the French Navy that wanted to focus on commerce raiding with torpedo boats and smaller ships, and the traditional battlefleet school, furthering the divisions that had wracked the French Navy since the early 1870s. Although the rising financial solvency of the French government in the late nineteenth century increased the available resources for military consumption (Sumida 1989, 7), France's naval budget trailed far behind Britain's. In 1906–7, while the British spent over £35 million on their navy, representing 55.1 percent of the total military expenditures, the French spent a little over £12 million on their navy, representing 26.2 percent of the total military expenditures (Gardiner, Gray, and Budzbon 1985, 190; Sumida 1989, 355).

Even if the French had spent the same percentage of their total military expenditures on their navy as the British, they still would have trailed significantly in total spending. Despite this shortfall, they still had a large enough naval budget to build the necessary battlefleet warfare technologies. Given this budget, plenty of information on British practices, and a strategic environment that favored a blue-water navy, one might expect the French to respond to British naval innovations by adopting.

But the organizational fissures in the French Navy stripped it of the capacity to successfully implement the innovation without delay. The Jeune École had attempted to shift the critical task of the French Navy from sea control and protecting trade with French colonies to commerce raiding and coastal defense. The Jeune École itself was a French attempt to counter British naval superiority by developing a cheaper way of conducting naval warfare that did not require large-scale mimicry of the British. Since the debate over the Jeune École lasted until the end of Pelletan's reign as minister of marine, though, this left the French Navy not just with a critical task ill suited to the innovation but also without a clear critical task at all. The victory of the advocates of battleship warfare within the French Navy left the French with a naval strategy ill suited to their actual position as well (Coutau-Begarie 1996, 61). This led to confused debates over the value of adopting the dreadnought model as the navy experienced internal disagreements over the purpose of the navy itself.

In one sense the French Navy arguably had an advantage in adopting the innovation because it did not have a stable naval bureaucracy attempting to maintain the status quo; its organizational age was young enough that veto points had not proliferated. Yet the lack of any stable naval bureaucracy, which reflected the chaos in the French military as a whole, created too much organizational confusion for adoption. While the French Navy led the world in experimentation in the 1880s at the height of the Jeune École, by the late 1890s internal conflicts had taken their toll on the desire of the government to spend money on costly experiments that did not seem to result in actual changes in French naval practice (Dahl 2005).

Thus, organizational constraints doomed the French attempt to fully adopt the innovation in the short-term, but their financial capacity allowed relatively early adoption of the technologies as they built the organizational capacity over time. More important, as a treaty partner with Great Britain, the first mover, France further bandwagoned with the English.

Difficulties with focusing the critical task of the navy, creating a stable naval bureaucracy, and freeing up resources for experimentation in the years prior to the innovation left the French Navy ill equipped to rapidly adopt the innovation in the short-term. France was unable to adopt the universally applicable elements of the Selborne Scheme. As French organizational constraints were reversed, France moved haltingly toward the adoption of battlefleet warfare, beginning with the authorization to construct dreadnought-quality battleships in 1909. Even when the French spent larger sums of money on their navy, though, the output was quite a bit less than would have been expected. "In his report upon the Navy Estimates in 1911 Paul Benazet adduced the striking fact that, while within 15 years Germany expended £100 million on her fleet and France £152 million, the German Fleet rose to second place among the world's navies and that of France sank to the fourth. He attributed this failure to negligence, neglect and careless maintenance routine" (Gardiner, Gray, and

Budzbon 1985, 190). Consistent with the country-specific predictions, France attempted to bandwagon with Great Britain more and more throughout the period, even attempting to formalize naval cooperation in the Mediterranean and ensure British protection of the French coastline in the event of a war. Given the victory of the battleship school of French naval thought, closer alliance with the British Navy became the only option because it was the only way that France could be protected by enough battleship assets to make the strategy practicable (Coutau-Begarie 1996, 61).

AUSTRO-HUNGARIAN EMPIRE

The early twentieth-century Austro-Hungarian Empire was mostly land-locked, with some ports in modern-day Italy (Trieste) as well as today's Slovenia and Croatia. Simple geographic reality would suggest that the country lacked substantial maritime interests. In 1913–14, foreign trade accounted for only 10 percent of Austro-Hungarian gross national product, in comparison with trade dependency hovering around 40 percent for Great Britain and Germany. Austro-Hungarian maritime trade did increase throughout the first decade and a half of the twentieth century, almost doubling from 1904 to 1913 to account for 20 percent of the total trade (Vego 1996, 32).

Despite low levels of maritime trade in comparison with the other European powers, the geostrategic interests of the Austro-Hungarian Empire almost all had naval elements, generating support for naval expansion. For example, while the Battle of Lissa between the Austro-Hungarian and Italian navies is generally considered a footnote to the Austro-Prussian War of 1866, it profoundly affected the subsequent peace negotiations between Prussia, Austria-Hungary, and Italy. An Italian victory could have enabled more expansionist Italian territorial claims and landlocked the Austro-Hungarian Empire. The Austro-Hungarian victory, led by Rear Admiral Wilhelm von Tegetthoff, weakened the Italian negotiating position and meant Italy had to settle for only taking Venetia (Marraro 1942; Sondhaus 1994, 1–4).

In the late nineteenth and early twentieth centuries, Austro-Hungarian naval interests revolved around two key issues: competition for control of the Adriatic with Italy and the prevention of hostile powers from dominating the Mediterranean (Vego 1996, 1–3). Italian claims over Trieste, the major Austro-Hungarian shipbuilding and trading port, as well as other Austro-Hungarian territory, meant the dual monarchy at the very least required a force capable of effective coastal defense. Increasing trade volumes between the dual monarchy and the Levant states (Egypt, Greece, and Turkey) generated interests in keeping the Mediterranean safe for trade. For example, Austria-Hungary opposed the opening of the Bosporus Straits to the Russian Navy on the grounds that it could destabilize the Mediterranean (Vego 1996).

At the outset of the crucial period of innovation, in May 1904, Austria-Hungary's navy was primarily configured for coastal defense. In all areas of naval power, from submarines to battleships, the Italian Navy outgunned Austria-Hungary's (Vego 1996, 19). While the dual monarchy possessed a few pre-dreadnoughts, the ships were poorly kept and staffed. The financial capacity of the Austro-Hungarian Empire was also severely constrained, limiting its ability to adopt the innovation. While the section above demonstrates the financial constraints facing the French Navy, the French naval budget in 1905 was over three times that of the dual monarchy, which spent 92.1 million crowns on the navy in 1905, in comparison with the equivalent 304 million crowns spent by the French (and the 795.7 million crowns spent by the British). By 1910, the relative spending gap had increased, with every major European power and the United States spending at least 200 million crowns a year on their navies, but Austria-Hungary only spending 85.1 million crowns. Given the financial constraints facing Austria-Hungary, the theory predicts there would be cost-induced delays in the adoption of the innovation.

The Austro-Hungarian Navy had a higher level of organizational strength than one might suspect, but was still handicapped by internal struggles and a lack of experimentation. It "considered battleships the backbone of its strength" (Vego 1996, 186). There was also internal confusion regarding the actual critical task of the Austrian Navy, as the dual coastal and seagoing orientation caused squabbles for resources between naval leaders (Sondhaus 1992; Vego 1996). Some wanted to focus on coastal trade, others on Italy, and still others on broader Mediterranean concerns. Thus, the critical task of the Austro-Hungarian Navy did not serve as a major barrier to adoption but it did not encourage adoption either.

Organizational age is a similar story. From the Battle of Lissa in 1866 up through World War I, the Austro-Hungarian Navy was remarkably stable, but micromanaged by the emperor and his relatives. The autocratic and personal naval rule (there were only five heads of the Austro-Hungarian Navy between 1866 and late 1913) kept down the number of organizational veto points, simply because authority was overcentralized in the hands of a few key naval leaders and members of the royal family. In contrast, experimentation was a serious weakness in the pre–World War I Austro-Hungarian Navy, in stark contrast to the 1870s' submarine and torpedo boat experiments promoted by the Austrian naval commander, Admiral Maximilian Daublebsky von Sterneck. As the Jeune École, Sterneck's inspiration, faded out of style, the "risky" Austro-Hungarian exercises concerning torpedo boats faded as well. Sterneck held on to power until 1897, probably delaying a shift by the Austro-Hungarian Navy toward preparing for the future (Sondhaus 1994, 97–103; Vego 1996). So while the organizational environment was not perfect for innovation adoption, the medium level of organizational capital required meant that, given appropriate financial resources, the Austro-Hungarian Navy would have the ability to adopt much of the innovation over the medium-term.

The results show the predicted early delay in the Austro-Hungarian response to the battlefleet innovation. The Austro-Hungarian Navy did not initially attempt to copy the *Dreadnought*. But after the crisis resulting from the Bosnian annexation issue in 1908 and the announcement that Italy, Austria-Hungary's competitor for influence in the Adriatic, would build dreadnought-style ships, Vice Admiral Montecuccoli, the head of the Austro-Hungarian Navy, personally authorized the construction of two dreadnought-quality ships in secret as he lobbied key political actors for funding to pay for them (Sondhaus 1994, 191–94). Montecuccoli's need to circumvent normal budgetary procedures and present the delegations with a fait accompli with regard to dreadnought-quality battleships highlights the importance of the financial limitations facing Austria-Hungary.

As a state allied to a likely adopter of the innovation (Germany), Austria-Hungary should also have attempted to bandwagon, according to the theory. This is especially true given the generally poor relations between Austria-Hungary and the first mover, Great Britain. Interestingly, as the British-German naval competition accelerated, relations between Great Britain and Austria-Hungary did not deteriorate as rapidly. When Austria-Hungary finally did begin building dreadnought-quality battleships in 1909 and 1910, the British assumed that Austria-Hungary's building program resulted from a request by German fleet guru Tirpitz, so they redrew war plans for the Mediterranean to reflect the new Austrian ships. In reality, despite the continuing presence of a formal treaty relationship between Austria-Hungary and Italy, it was the Bosnia crisis and concern about the Italian battleship buildup that drove the shift in Austria-Hungary's policy. While cooperation between Austria-Hungary and Germany was likely in the event of a war, the idea that Austro-Hungarian naval constructions plans occurred as a result of Germany was simply incorrect (Sondhaus 1994, 191–93 202–3).

JAPAN

In the aftermath of World War II and over sixty years of Japan's unique brand of antimilitarism, it is easy to forget Japan's impressive rise to global naval prominence. The modern Japanese state did not even emerge until the Meiji Restoration in 1868. At this point Japan had little to no naval power, excluding some older vessels staffed by enthusiastic but poorly trained troops. Fewer than thirty years later, in 1894–95, Japanese naval forces defeated the best the Chinese had to offer. A mere ten years after the conflict with China, Japanese naval forces dealt a decisive blow to the Russian Navy at Tsushima, terminating the large-scale Russian naval presence in East Asia. This left Japan, now an alliance partner of the British, as the predominant naval power in the Pacific.[24]

---

[24] Naval Necessities, vol. 2, 1905, 153–59, box 1, paper 7, Admiralty 2-15-1906, box 2, paper 13, Crease [Fisher] Papers.

The speed with which Japan achieved this goal—dominating the Pacific Ocean fifty-two years after Commodore Matthew Perry visited in 1853—is a remarkable achievement that has faded in historical memory due to Japanese actions in subsequent decades.

The Japanese Navy from the beginning sought to learn from foreign powers. Japan conspicuously adopted the British Navy as a "model" for its own, and sought, through officer exchanges and ship orders, to emulate the British way of training and fighting as well as British ships (Falk 1936). The Japanese naval academy, Eta Jima, was based on the British naval academy at Dartmouth, and Naval Commander Richard Tracey was sent from Britain to help establish the academy (Gardiner, Gray, and Budzbon 1985, 222).[25]

The Japanese Navy played a major role in the 1894–1895 Sino-Japanese war with its defeat of the Chinese Navy at the battle of Yalu River, helping secure larger future funding for the navy. This was crucial, as the Japanese Army and Navy competed heavily for resources during this period (Gardiner, Gray, and Budzbon 1985, 221; Nish 1996, 78).[26] Ultimately, the army recognized that successful land conquest depended on sea control for landings and stable supply lines. U.S. president Theodore Roosevelt then heralded Japan's naval victory over Russia in the Battle of Tsushima as the Trafalgar of its time—if not more important—and along with the alliance with Great Britain signed in 1902, the victory established Japan as the dominant naval power in the Pacific (Hendrix 2005).

Before the victory at Tsushima, the Japanese Navy had already recognized the value of large-caliber armaments. Experience in battle with the Russians showed that six-inch shells were like "pea shooters," according to the British naval attaché on sight, and the larger the caliber, the bigger the impact. This was especially true for longer-distance engagements, meaning the accuracy of smaller-caliber weapons no longer made them superior to larger-caliber ones. The navy understood this, and as a result the Satsuma-class battleship, planned for construction around the same time as the *Dreadnought*, was supposed to have a uniform large-caliber armament. Unfortunately, Japanese dependency on guns imported from Britain meant that cost overruns on the British side forced a design change toward mixed-caliber armaments. Conceptually, however, the Japanese were already on board with the *Dreadnought* when its existence was revealed to the world (Breyer 1973, 331; Falk 1936).

The financial intensity of dreadnought ships overwhelmed Japanese production capacity in the short-term, though. The cost of the Russo-Japanese war

---

[25] Japan also had close ties with the French Navy in the 1870s and 1880s, purchasing cruisers and torpedo boats from the French, but the initial influence on the Japanese Navy came from Great Britain (Perry 1966, 309).

[26] For more on the role of the Japanese military, see Ienaga 1978. For a critique of Japanese navalism in the 1930s, see Kawakami 1932.

and the subsequent global economic recession imposed severe constraints on Japan's naval procurement budget. The first dreadnought-style ships were not laid down until the Settsu class in 1909, and the Settsu class was only nominally a dreadnought. Still, in the years immediately preceding World War I Japan caught up in warship quality. The Fuso class, indigenously produced but partially from British designs, was clearly a super-dreadnought. Meanwhile, the Kongo-class battle cruisers, some produced in England and others in Japan, achieved significant speed increases and obviously accepted Fisher's vision of the battle cruiser (Gardiner, Gray, and Budzbon 1985, 223, 234). So the financial costs of the battlefleet innovation led to some delays in technological adoption, but Japan's high level of tacit knowledge meant that the technology was adopted quickly once funding was available.

The Japanese Navy had a high level of organizational capital, which would also predict its adoption of the battlefleet innovation. While the Japanese Navy was originally designed around coastal defense, driving its interest in French cruisers and torpedo boats in the 1870s and 1880s, by the 1890s it came to view the destruction of enemy fleets as another element of its critical task. Japanese naval historian Aoki (1983, 355, cited in Nish 1996, 78) writes,

> In the view of naval officers at the time, the object of holding sea power was not for the sake of protecting lines of communication for the sake of commerce, as Mahan considered, but was for the sake of destroying enemy fleets which came to attack Japan, and protecting Japan's coasts.

This made the critical task of the Japanese Navy well suited to adopt battlefleet warfare and probably would have made it more possible for its adoption of flotilla defense as well, since Japan's acceptance of some of the ideas of the Jeune École showed its capacity to think beyond the battlefleet. Even so, the battlefleet dominated Japanese naval strategy, especially after 1905 as the navy turned to Mahan to provide the theoretical justification for developing a navy capable of winning a modern decisive battle (Evans and Peattie 1997, 24). The Russo-Japanese War had demonstrated the possibility and importance of decisive naval battles. The need for larger and more expensive ships also helped the navy in budgetary debates with the army (Evans and Peattie 1997, 129–30, 149).

One area where the Japanese Navy trailed some of its Western counterparts was a commitment to experimentation. Yet the Japanese Navy's low organizational age may have helped it overcome that weakness. As a new naval power without a serious back history in the age of sail, the Japanese Navy lacked the built-in veto point and bureaucratic barriers that might prevent rapid adoption of new technological and strategic innovations. There were no old admirals arguing that things should be done the way they "used to be done," and no armaments industry with hundreds of years of tradition in the way it produced ships. This ironically made the Japanese Navy a "healthy" bureaucratic structure and helps explain how it ended up adopting the innovation. As Michael Barnhart (1999, 211) observes,

> It will come as a surprise to some readers that Japan lacked a maritime tradition . . . . in 1853. . . . Perhaps this was a blessing, as Japan (actually the daimyo of Satsuma, the true pioneer of modern Japan's naval forces) was able to begin with a clean sheet and an open mind.

Given Japan's treaty relationship with Great Britain beginning in 1902, adoption-capacity theory would predict a slight delay in the Japanese adoption of the battlefleet and Japanese movement toward bandwagoning with Great Britain. This only partially occurred. Japan actually thought seriously about not getting involved at all in World War I, though it did eventually join the entente powers and participate in several World War I naval operations (Nish 1996, 79). One explanation for why bandwagoning did not occur is that the Japanese Navy had already made so much progress toward adoption prior to the introduction of the *Dreadnought* that this was more a case of simultaneous development than diffusion. Bandwagoning was not necessary even though Japan and Britain continued to ally. Then again, it is also possible to argue that Japan and Great Britain were not *real* allies; Japan still worried about British expansion in Asia despite their cordial relationship. Japan also had independent strategic motives to adopt the innovation deriving from concern about the growth of the U.S. Navy and the German fleet (Admiralty 2-15-1906, box 2, paper 13, Crease [Fisher] Papers, 10–12; Evans and Peattie 1997, 166–67).

Conclusion

It is possible to separate the key naval innovations of the late nineteenth and early twentieth centuries into two categories: the battlefleet innovation represented by the mighty *Dreadnought*, and the flotilla innovation symbolized by the torpedo boat and submarine. The spread of the former occurred in line with the predictions of adoption-capacity theory. Key organizational and financial constraints, in combination with potential benefits from alliance relationships, predicted the timing of technological adoption and the extent of organizational adoption by the major powers. With degrees of delay, major powers adopted the innovation, but while some of the technologies spread even beyond the major powers, few minor powers were able to adopt the organizational framework of battlefleet warfare. The implications for the balance of power also functioned as expected. The innovation increased asymmetrical relations between major and minor powers, and more important, gave the first mover, Great Britain, short-lived but significant breathing space in the race for naval superiority prior to World War I. The British failure to adopt flotilla defense also provides more evidence in favor of adoption-capacity theory, though applied to a case of innovation, rather than diffusion, which this book does not explore in detail. Flotilla defense was a different story; it disrupted several hundred years of thinking about how to conduct naval warfare in a way that the *Dreadnought* and *Invincible* never did. The proliferation of veto points in the Admiralty, in combination with the massive gap between the critical task of the British Navy and the flotilla innovation, allowed opponents to frustrate attempts to implement the innovation at nearly every step, delaying adoption long enough that a major war intervened and the innovation became a forgotten cautionary tale in British naval history.

In the end, the debut of the HMS *Dreadnought* influenced the global naval balance of power. The medium levels of financial capacity and organizational capital required for adoption led to a dissemination of the technological components of the innovation at a steady clip. The adoption of the completed innovation, including its organizational components, lagged as it does for nearly all innovations, but not nearly as much as in the case of an innovation like carrier warfare, where the enormous organizational requirements overwhelmed most naval powers.

As Fisher believed, the introduction of a ship like the *Dreadnought*, an all-big-gun ship that also tried to maximize speed and armor, was probably inevitable. The process by which it happened, however, is worth noting. The huge difference between pre-dreadnought and dreadnought ships created a naval power gap that actually helped new naval powers like Germany gain on Great Britain after an initial period of greater inferiority, as they were not faced with the challenge of making over or abandoning an old and out-of-date fleet. Japan's high level of organizational capital helped it build a navy that would soon have the ability to equal the British Navy—before passing it in the 1930s at the dawn of the carrier age. The battlefleet innovation also helped speed the relative demise of the French Navy, which managed to last as a major naval power only through the beginning of World War II. As the primary competitor to the British Navy throughout the nineteenth century, France would never again challenge the British Navy for control of the oceans—or even control of the areas surrounding France. Together, these examples illustrate how even a short-lived innovation in the strategy of war fighting can have a lasting impact on the global balance of power.

Chapter 6

SUICIDE TERRORISM

IN THE MID-1990s, after the first World Trade Center attack, Osama Bin Laden apparently made an important decision with major repercussions for U.S. strategy. Up until then, the burgeoning terrorist group now known as Al Qaeda had played a major role in Salafi jihadi terrorist operations around the world, but its involvement was mostly behind the scenes. Al Qaeda provided financing for operations, trained fighters from affiliated groups, and smuggled weapons to sympathetic parties. Yet Bin Laden, Al Qaeda's leader, determined that it was time for Al Qaeda itself to engage in a major attack and step out of the shadows. When planning began for the operation that was to become the East African embassy bombings of 1998, Bin Laden sent some of Al Qaeda's top military commanders and operatives, including some in the Kenya cell, to Hezbollah to learn from one of the most successful terrorist groups of the last twenty years. Even though Bin Laden's Sunni Salafi beliefs led him to clear theological disagreements with the Shia-affiliated Hezbollah, and Hezbollah had not actually conducted a suicide attack in years, Bin Laden considered them the experts and sent his people to learn. Furthermore, Bin Laden purportedly told his operatives to specifically study the 1983 U.S. Marine barracks suicide bombing by Hezbollah. His operatives went, took careful notes, and returned with the operational concepts and knowledge necessary for the embassy bombings in 1998.[1]

Though focused on nonstate actors, the story illustrates three critical concepts of adoption-capacity theory. First, sometimes technical expertise is not enough to adopt an innovation. Even though Al Qaeda had money, committed members, and weapons, it needed to send its members to Hezbollah, a suicide terrorism innovator, to pick up the tacit knowledge necessary to conduct its own suicide operations. Second, organizational capital matters. Al Qaeda's lack of a prior operational history made it extremely flexible when it came to designing the embassy bombings. Without an operational past that caused Al Qaeda to privilege certain attack strategies, it was easier to branch into a new area of operations like suicide bombing. Third, it is impossible to tell the story of how a type of military power matters without understanding how it spreads.

---

[1] Nearly all of the evidence from this story is taken from *The 9/11 Commission Report*. The section from which this story is told cites multiple U.S. intelligence briefs and testimony by group members in court cases (National Commission on Terrorist Attacks upon the United States 2004, 67–68, 470–71).

The connection between Al Qaeda and Hezbollah became a crucial node in the spread of suicide terrorism around the world—a link connecting a key innovator in the 1980s, Hezbollah, to the primary exporter of knowledge about suicide terrorism from the mid-1990s to the present, Al Qaeda. These three lessons from the theory are pivotal to predicting the spread of military power, whether conventional military innovations or activities of nonstate actors.

Studies of terrorism in general and suicide terrorism in particular tend to view terrorist groups independently. Work by Robert Pape (2005, 45–47) on suicide terrorism focuses on foreign occupation and religious differences between the terrorist group and the perceived occupying state as driving suicide bombing. Similarly, Mia Bloom's market share and outbidding theory (2005) presumes groups make decisions about the adoption of suicide terrorism based on their need to compete for influence with other local terrorist groups. While each author mentions the mass of interrelationships between many terrorist groups, they generally assume the independence of each observation in the "data" of suicide terrorist attacks, although Bloom explicitly recognizes linkages between groups within disputes like the Israeli-Palestinian conflict.

But what if the propensity for a terrorist group to adopt suicide tactics depends in part on its external linkages, and the relationship between the organizational capabilities required to adopt the innovation and the organizational capabilities of the group? If diffusion processes influence who adopts strategies and when, any theory of nonstate innovation that ignores diffusion risks missing critical factors that affect behavior.

The evidence presented below shows that suicide terrorism requires low levels of financial intensity but high levels of organizational capital to implement. There is a direct relationship between organizational capital and the adoption of suicide terrorism. Terrorist groups with preexisting operational profiles before the era of suicide terrorism began in the early 1980s had extreme difficulty adopting the innovation, which makes sense given their generally low levels of organizational capital and relatively high organizational ages. Leading pre-1980s' groups like the PLO, the PIRA, and ETA all failed to adopt suicide terrorism in the short- and medium-term. It was not until the midst of the Al Aqsa Intifada that the PLO adopted suicide terrorism, despite clear strategic incentives to adopt in previous years.

The interaction of organizational capital with religion, a key transmission mechanism of the tacit knowledge necessary to adopt, explains both which groups are most likely to adopt suicide terror and which are not. Networks of religiously motivated groups, through the direct diffusion of knowledge from group to group and demonstration effects that influenced nonreligiously motivated groups, distributed suicide terrorism around the world. For religiously motivated groups in particular, there is a direct relationship between organizational age and the probability of adoption. Young groups are likely to adopt, but the probability of adoption drops sharply over time.

Defining Suicide Terrorism

Defining terrorism and suicide bombing remains a controversial task. While some definitions focus on the targeting of civilians as the essence of terrorism, others concentrate on the distinction between state and nonstate actors (Hoffman 1998, 15). While the Zionist Stern Gang in the 1940s publicly referred to itself as terrorist, most groups thought of as terrorist today instead refer to themselves as liberation or freedom fighters (McCauley 1991, 127). Arguments about defining terrorism range from the belief that one person's freedom fighter is another person's terrorist to the idea that the terrorism label is a rhetorical tactic designed to discredit groups and political goals to the question of whether the firebombing of Dresden or the nuclear bombing of Hiroshima and Nagasaki represented acts of terrorism (Hoffman 1998; Wilkinson 1986). Recognizing the inherent contingency in the definition, this chapter defines terrorism as violence and/or the threat of violence by nonstate actors, potentially funded by a state and designed to influence government policy through a variety of media, including public opinion.

It is easier to lay out what distinguishes suicide from nonsuicide bombing. Suicide bombing acts are violent attacks designed to kill others through an act that must include the death of the terrorist or terrorists. This definition excludes individual suicides because an individual suicide does not kill others. High-risk military missions, sometimes called suicide missions, are also distinct from suicide bombings because the death of the soldiers is not the means by which the mission is accomplished; death is just highly likely due to complicating factors. For an act to be defined as a suicide bombing, the death of the bomber must be how the act is completed (Kondaki 2001).

Existing Suicide Terrorism Literature

Most research on suicide terrorism has focused on what motivates the groups and individuals that authorize and conduct these acts. What makes individuals decide not just that they are willing to die fighting for a cause but that they are willing to die deliberately as a means of inflicting harm on others? After decades of research, it seems that suicide terrorists, on balance, are not generally mentally ill or afflicted with some sort of psychological condition. Individual-level motivations for volunteering include revenge against governments that killed loved ones, despair due to hopeless economic or social conditions, social pressure, or other personal crises (Fearon and Laitin 2003; Gordon 2002; Lester, Yang, and Lindsay 2004; Sageman 2004; Weinberg and Pedahzur 2004). Alan Krueger and Jitka Maleckova find no relationship between economic distress and support for terrorism, while other theories focus on the proportion of young men in a given country, and whether the conflict involves territory (Hassner 2003;

Hudson and Den Boer 2002; Krueger and Maleckova 2003; Toft 2003). Marc Sageman (2004) casts doubt on any particular individual-level behavioral pattern by demonstrating, through a study of several hundred individual terrorists, the lack of a common background or enabling condition.

Pape argues that when territorial occupiers are democratic states, meaning their policies are sensitive to changes in domestic public opinion, and they are of a different religion than the group being occupied, suicide terrorism by nonstate actors becomes more likely.[2] Occupation is the critical determinant of whether or not suicide bombing occurs (Pape 2005, 21). For Pape, groups rationally utilize the tactic most likely to lead to success—and since suicide attacks seem to empirically succeed in driving out occupying forces, groups use suicide terrorism.

Much like Pape, Bloom also views the adoption of suicide terror tactics by terrorist groups as rationally based on cost-benefit analysis. Instead of foreign occupation, however, Bloom argues that internal competitions for influence within oppressed communities create incentives for groups to seize "market share" of public opinion by "outbidding" each other through demonstrating higher levels of dedication to the cause. Suicide terrorism signals intense commitment, since by definition it involves the death of a group member, meaning that in supportive communities, suicide terrorism can increase approval levels for the group perpetrating the bombings (Bloom 2005, 78–79). This creates internal political incentives for groups to adopt suicide bombing regardless of the instrumental impact on the adversary.

Assaf Moghadam disagrees with both Pape and Bloom, writing that suicide terrorism has become a globalized phenomenon. The transnational nature of the demands by jihadi groups means local bargaining or occupation explanations have inherent limits; they cannot account for the newest and most deadly variants of suicide terrorism (Moghadam 2006).[3]

A limited amount of the existing terrorism literature has viewed it from a contagion perspective. Focusing on the spread of terrorism within Latin America and then to Europe in the 1960s and 1970s, Manus Midlarsky, Martha Crenshaw, and Fumihiko Yoshida contend that diffusion depends in large part on the imitation of one group by another without direct contact or interaction. They find that an inverse diplomatic hierarchy governs the spread of international terrorism—terrorist groups in more developed countries model the efforts of terrorist groups in less developed countries (Midlarsky, Crenshaw, and Yoshida 1980). In contrast, Edward Heyman and Edward Mickolus (1980) state that terrorism spreads based on the status of the terrorist group, not the diplomatic status of the country in which the group operates.

---

[2] The goal is to influence the government and people by inflicting costs.

[3] Indeed, in many ways Moghadam's work is a natural leaping-off point for this chapter. For other critiques of Pape, see Jackson Wade and Reiter (2007); Bloom 2005, 84–85.

Bruce Hoffman (1998) has detailed the extremely high number of interlinkages between terrorist groups over time—linkages that fuel the diffusion of tactics between groups.

Global terrorist hubs like Lebanese and Libyan training camps in the 1970s and 1980s, and Sudan and Afghanistan in the 1990s, have served as clearing-houses for terrorist group coordination. The training camps in the 1970s and 1980s in the greater Middle East, mostly run by the PLO, hosted terrorists from around the world, even including some operatives from North Korea (Bloom 2005, 121). While most joint training has occurred between groups with similar ideological outlooks, instances like the PIRA's training of Revolutionary Armed Forces of Colombia members and ETA members' training with the PLO show that terrorist group cooperation has extended beyond traditional ideological lines (Murphy 2004). Many scholars have noted the cooperation between Hamas, Islamic Jihad, and Hezbollah in the early 1990s that likely included the diffusion of knowledge related to Hezbollah's suicide terrorism experience. Bloom points to the training of jihadis in the Afghan-Soviet War in the 1980s and training camps in Afghanistan in the late 1990s as a critical nexus through which suicide terror has spread in subsequent years (Bloom 2005, 138).[4]

Suicide terrorism can diffuse through both direct and indirect means. The direct diffusion of suicide terrorism occurs when groups physically coordinate and train together, and knowledge is transferred from one group to another. Hezbollah operatives training Hamas operatives after their expulsion to Lebanon in 1992 are an example of direct diffusion. Indirect diffusion occurs when one group reads or otherwise learns about the exploits of another group, and then attempts to model those exploits. When reports of the suicide vest created by the Liberation Tigers of Tamil Eelam (LTTE) in Sri Lanka inspired similar tactics by Hamas, this was indirect diffusion.

SUICIDE TERRORISM AS A MAJOR MILITARY INNOVATION

Terrorist Groups and Military Organizations

It may seem counterintuitive to compare military organizations run by nation-states with nonstate actors. But the financial and organizational constraints outlined by adoption-capacity theory are pertinent to terrorist groups as well as military organizations. Terrorist groups, like conventional militaries, face resource constraints that influence all aspects of their planning process, from how often they attack—the operational tempo—to who they plan to attack

---

[4] Bloom lays out many of these linkages between groups and even discusses why the PIRA never adopted suicide terrorism—a topic dealt with in more detail below. Nevertheless, Bloom (2005, 138) does not outline a theory of how and why suicide terror diffuses, though she implicitly argues that direct connections with groups that conduct suicide terrorism make the diffusion of the tactic to a new group much more likely than if direct ties do not exist.

and how they plan to conduct an attack. The availability of resources influences the types of equipment available, such as the types of bombs or small arms that a group can build or purchase. Financial resources also influence the ability of a group to send potential actors off for training at external sites or buy safe houses to shield group activities from the government.[5]

Terrorist groups also face organizational constraints. Once terrorist groups form, plan, and conduct operations, they develop at least tacit bureaucracies and hierarchies—and sometimes even explicit bureaucracies and hierarchies—that rival state-based military organizations. Group members gain or lose prestige depending on whether their ideas succeed or fail, and subunits may gain or lose prestige based on their ability to plan and conduct specific types of operations. Just like businesses and military organizations, terrorist groups have traditionally developed expertise at particular tasks; instead of producing widgets or fighting tank battles, terrorist groups learn to assault military bases, hijack airplanes, or build remotely detonated explosives. Once terrorist groups develop organizational expertise in particular types of attacks, extending the idea of organizational age predicts that they should develop veto points and bureaucratic obstacles to change. While the institutions may be more informal, research by Douglass North (1981) on economic institutions and Wilson (1989) on bureaucracies proves that informal "rules" and ways of doing business also function as informal institutions that regulate behavior.

Is Suicide Terrorism a Major Military Innovation?

While most writing on suicide terrorism has not linked the phenomenon to other military innovations, according to the definition of a military innovation, suicide terrorism fits as a case for analysis. Explosives—the technological elements that form the building blocks of nearly all suicide terrorist attacks—were invented and perfected in the late nineteenth and early twentieth centuries (Elster 2005, 234). Some smaller technological changes then combined in ways that made most forms of suicide terrorism more possible beginning in the 1980s. The decreasing size of explosive charges allowed terrorist groups to generate a larger impact from less material. This meant that individuals literally carrying bombs on themselves were even more lethal than before and that the relative impact of suicide car bombs also increased. But the importance of these changes should not be overstated. It is the organizational consequences of adoption rather than the purchasing of the hardware that really makes suicide terrorism a military innovation.

---

[5] Terrorist groups often utilize different methods of extracting resources, however, from state sponsors to criminal activities to direct fund-raising. This may also influence the propensity of groups to use certain strategies and is an important avenue for future research. I am grateful to an anonymous reviewer for making this point clear.

Suicide terrorism, organizationally, is an attempt by a group to circumvent asymmetrical weakness by using members of the group themselves as part of the delivery mechanism for attacks (Merari 1990; Pape 2003). It substitutes people (and sometimes people in cars or planes) for artillery, missiles, and other weapons. Suicide terrorism is also an attempt to circumvent the barriers to assassination and attack presented by modern security screening. It is simply much more difficult to approach a national leader today than in the time of the Assassins in the Middle East, and it is difficult to drive a truck full of explosives up to the White House, get out of the vehicle, and then explode it from afar. So suicide terrorism, in one form, is a potential organizational response to the challenge of gaining access to and destroying particular types of targets.

Creating a suicide terror attack requires shifting how a group does business. The type of training operatives get for suicide attacks is different than the training they get for other types of attacks. Terrorist training at the tactical level has traditionally placed at least some emphasis on trying to evade capture after an attack and how to handle interrogation if capture occurs. The evidence from the Palestinian and LTTE cases suggests that ideological training concerning the justness of the cause and the action is substituted in the suicide terrorism case. Suicide terrorism also means that group members chosen to conduct an operation will undertake new and different tasks, and operational success necessitates the permanent loss of the operative to the terrorist group. This means that terrorist groups have to think about shifting the way they recruit members and select which members will conduct attacks. Someone with extensive experience in evading authorities and launching mortars at government troops might be the perfect person for a nonsuicide mission, but not a suicide bombing mission, both because their primary training is in another form of attack and because loss of that operative entails a permanent human capital loss for the organization.

While the diversity of groups using suicide terrorism means that each group is different, each group that has used suicide terrorism has had to change its recruitment practices. When suicide attacks are mostly used against hard targets, meaning the goal of the attack is an instrumental on-the-ground military accomplishment, the group needs an extremely high attack success rate. One solution is to create special ranks of group members trained to conduct suicide attacks. For example, after the LTTE recruited new members, it sent the best to specialized training where they attempt to become Black Tigers, members of the group from which the LTTE selected suicide attackers (Hopgood 2005; Jackson Wade and Reiter 2007, 63–64).

If suicide attacks are mostly used against softer civilian targets, the success rate requirements are likely lower, both in terms of the casualties per operation, and whether or not the operation succeeds at all. For instance, while both Hamas and Islamic Jihad initially used highly trained long-term operatives to conduct attacks, by the Al-Aqsa Intifada, beginning in 2000, both groups shifted to recruiting soft supporters from the community for specific suicide terror operations and training them for short periods of time, mostly for ideological reinforcement (Pedahzur 2005, 169). This avoided risking the human capital of trained members. Either way, suicide terror operations require different recruiting and training than nonsuicide terror operations.

The combination of the use of technology in a different fashion, in an operation that necessitates killing the carrier of the weapon in order to damage opponents, and the different recruiting and training methods required to conduct suicide attacks means that suicide terrorism is a military innovation. Suicide terrorism is unparalleled in its ability to inflict casualties relative to the cost of the attack. A suicide attack, according to evidence presented by Scott Atran and captured by the Israeli government, costs about $150 (Atran 2003, 1537; Israeli Defense Forces pokesperson 2002; Jones 2003, 281–82).[6] The average number killed and wounded in suicide attacks tends to exceed that of other types of terrorism, though there is variability ranging from the highly destructive attacks of Al Qaeda and Hezbollah to the less destructive ones of Hamas and the Kurdistan Workers' Party (Ricolfi 2005, 98). Evaluating terrorist attacks from 1980 to 2001, Pape (2003, 5) finds that suicide attacks composed 3 percent of the total number but accounted for almost 48 percent of the deaths. Suicide terrorism is also more accurate and controllable than other forms of attacks. The terrorist or an external controller can decide exactly when to detonate the bomb to maximize or minimize casualties depending on the situation, or change locations to alter the desired impact. It has been said that "the suicide terrorist is the ultimate smart bomb" (Hoffman 2003, 40; Pedahzur 2005).

The success of terrorist groups in achieving their goals, in a historical sense, is quite low. While not all innovations are successful, suicide terrorism has given terrorist groups much more success than nonsuicide operations.[7] The reason that groups resort to terrorism in the first place is often because they lack the conventional military capabilities to defeat a government and/or because they lack the public support to mobilize the population to take up arms against the government. Usually a combination of both factors is present. Pape (2003, 9) argues that suicide terrorism, historically, has carried a success rate above 50 percent. Moghadam (2006, 713) disagrees, saying that the success rate is only 24 percent.[8] The reality is less important than the perception among

---

[6] All of the peripheral intelligence and postattack costs, like payments to families, generally occur for other types of attacks as well so they are not unique costs for suicide terrorism.

[7] It is important not to view all suicide bombing campaigns as the same. Luca Ricolfi (2005, 105, 114–16) points out that given the multiplicity of motivations for suicide terrorism, it is crucial to theorize in ways that recognize these differences and do not focus on one-size-fits-all explanations.

[8] In another review of Pape, Max Abrahms (2006, 43) maintains that at least one set of terrorist groups, those on the U.S. State Department terrorist list, only succeed 7 percent of the time.

terrorist groups about the success of the tactic. The perceived substantial success of the first mover, Hezbollah, in using suicide attacks to drive the West and Israel out of Lebanon, and a prominent adopter, the LTTE, in using suicide attacks to bring Sri Lanka to the negotiating table must send a relatively clear signal to other terrorist groups that suicide bombing works. There are also potential benefits to suicide terrorism related to its symbolism as a demonstration of the group's commitment to its cause, and the use of the tactic to "outbid" other resistance groups and build credibility with the local population (Bloom 2005; Hoffman and McCormick 2004; Kramer 1991).

The Debut of Suicide Terrorism

The human as a bomb is not an entirely new method of employing military force; late nineteenth- and early twentieth-century anarchists as well as Japanese kamikaze pilots both engaged in suicide bombing according to the definition above.[9] Yet the Lebanon bombings in the early 1980s signaled a new era of suicidal military activity.

In the wake of the Lebanese civil war, the Shiite population in Lebanon concentrated in the south and around Beirut. Several groups, most prominently Amal, sprang up to help defend Shiite interests in the midst of the sectarian strife. In 1982, the Israeli occupation, continued Maronite-Palestinian violence, and the deployment of Western troops caused a splinter within Amal. The more radical elements, which sought to establish an Islamic state in Lebanon, moved to the Bekka Valley and joined forces with over a thousand Iranian Revolutionary Guards sent by Ayatollah Ruhollah Khomeini. The group took over a Lebanese army fortress and the surrounding territory, naming itself Hezbollah, or "Party of God" (Kramer 1990). On November 11, 1982, Hezbollah launched its first suicide attack: a bombing near an Israeli military installation in Tyre (Pedahzur 2005, 47).

Given that Iranian delegates helped with operational planning in Hezbollah's early years, it is possible that the Iranian use of human waves, or adolescents sent to the battlefront to literally walk across minefields in de-mining operations and/or attempt to overwhelm Iraqi forces en masse, provided the operational inspiration for suicide terror in Lebanon, especially given the Iranian justification of these human waves as Islamic martyrs (Pedahzur 2005, 4; Ricolfi 2005, 87). Hezbollah also viewed itself as involved in a struggle for the future of Islam and the people of Lebanon; its members had an obligation

---

[9] Early anarchists had nonhierarchical organizations, but these had limits for ideological reasons. These groups had nebulous political goals, but very rarely specific goals. The kamikazes were conducted by a nation-state, not a nonstate actor, and were a weapon of last resort. While the kamikazes do not appear to have inspired the current generation of suicide terrorists, their actions certainly fit the definition of suicide attacks.

according to their interpretation of Islam to practice jihad and fight to expel the invaders. So-called self-martyring operations were justified by clerics on the grounds that if the martyr was genuinely devoted and the operation caused sufficient destruction of the opposition, it was a theologically and pragmatically useful tactic (Kramer 1991).[10] Martin Kramer cites statements by the spiritual leader of Hezbollah, Sayyid Muhammad Husayn Fadlallah, who said that "the Muslims believe that you struggle by transforming yourself into a living bomb like you struggle with a gun in your hand. There is no difference between dying with a gun in your hand or exploding yourself." Fadlallah also maintained, "What is the difference between setting out for battle knowing you will die *after* killing ten [of the enemy], and setting out to the field to kill ten and knowing you will die *while* killing them?" (Kramer 1991).

The suicide-bombing period in Lebanon and the competition with Amal gave Hezbollah global notoriety that far outstripped its actual in-country influence at the time (Kramer 1993). The nonstate nature of the act, the casualties from the initial demonstrations, and the public nature of the media coverage make the early Lebanon bombings the appropriate point at which the innovation should be considered "mature."[11]

If suicide terrorism is an effective military innovation whose power has been demonstrated in conflict, it makes sense to ask the same questions about suicide bombing that are asked about other military innovations. When examining a conventional innovation, analysts tend to inquire, "Why didn't country X adopt this military innovation?" When looking at suicide bombings, however, the question has generally been "Why *did* group X adopt suicide terrorism?" If suicide terrorism is a military innovation, it makes sense to think about adoption as a strategic choice, and evaluate the factors that make both adoption and nonadoption likely (Kalyvas and Sánchez-Cuenca 2005, 209).[12] Given a set of terrorist groups in the international system, once they learn about suicide terrorism, they have to decide whether to try and adopt the innovation. The decision may or may not require additional learning or contacts with more experienced groups, or could mean deciding against adopting and attempting to achieve their goals through other means. If suicide terrorism is so effective compared to other types of terrorism, what has stopped most terrorist groups in most time periods since the Lebanon campaigns from using it? Instead of beginning by trying to explain why Hamas or Al Qaeda

[10] Note the functionalist justification for suicide bombing in this case. When the deaths per attack declined later in the 1980s, Shiite clerics in Lebanon banned suicide operations unless they had a high probability of dealing a major blow to the enemy (Kramer 1991).

[11] Even before the U.S. embassy bombing on April 18, 1983, sixty-one died in a bombing on December 15, 1981, at the Iraqi embassy in Beirut and eighty-five died in a bombing on November 11, 1982, at an Israeli military headquarters in Tyre (Ricolfi 2005, 80).

[12] Bloom (2005, 76) also frames the question is terms of why some groups use suicide terrorism while others do not.

uses suicide bombing, it is more useful to figure out why the vast majority of terrorist groups do not.

Predicting the Spread of Suicide Terrorism

This chapter is different from the other empirical ones in several ways. Since the diffusion of the innovation is ongoing, the conclusions drawn here are tentative. Unlike nation-states, since terrorist groups exist on the basis of their violent opposition to a government or other group, a shift to neutrality or "hiding" is not an option. Additionally, innovations adopted by a nation-state are generally designed explicitly for use against other nation-states, meaning strategic choices in response to innovations are, in theory, directly competitive.

Even if they exist in a highly competitive environment, though, terrorist groups might also coordinate at times.[13] In the nation-state context, alliances can theoretically allow states to substitute paying the cost of adoption for paying the cost of allying through a reduction in their freedom of action. Alliances can also sometimes allow states to more quickly acquire the technology and knowledge necessary for adoption from an alliance partner. For terrorist groups, the small sizes of most groups and their independent goals mean the protection function of alliances is usually not possible. Still, direct cooperation for the purpose of exchanging information about best practices can and does occur, and might significantly influence the probability of adoption.[14] While less formal than "epistemic communities," shared beliefs about effectiveness as well as the way to weigh costs and benefits could shape how a terrorist group makes decisions about whether to adopt an innovation (Haas 1992). Direct or indirect contacts between groups could drive a learning process. The learning process could even look like emulation if preexisting factors such as ethnicity, religion, language, or other things serve as the locus for diffusion (Gray 1973).[15] The mechanism for diffusion becomes the direct transmission of information from group to group or mimicry through vicarious learning.

For religion in particular, some scholars argue that the religious orientation of many new terrorist groups and the supernatural rewards offered for participation in acts like suicide attacks over the last few decades make religion a potential locus of adoption (Asal and Rethemeyer 2008; Benjamin

---

[13] Terrorist groups in the same region do sometimes compete for followers, meaning they may adopt tactics to boost their relative standing in the population or even on occasion attack each other (Bloom 2005). Nation-states could also adopt for other reasons, but the core purpose of the innovation is generally military.

[14] See figure 6.1 below. Alliances might also facilitate the diffusion of an innovation at a lower cost.

[15] For example, Elkins and Simmons (2004) find that cultural similarity matters for predicting financial policy diffusion.

and Simon 2002; Hoffman 1998). The intense personal and group-based factors driving religiously motivated groups could make them especially likely to adopt on exposure from similar groups. The transnational character of religious motivations also potentially makes religious groups candidates for networklike diffusion effects. As explained above, though, suicide attacks diffused from Hezbollah to Al Qaeda despite strong conflicts in their theological perspectives, though both are Islamic. The assertion here does not depend on the unique characteristics of any particular religion but rather the ability of religion to serve as a coordination vehicle for like-minded groups.[16]

Regardless of the mechanism for diffusion, what does a theory about the diffusion of military power developed to explain the behavior of nation-states have to say about suicide terrorism? This chapter posits that it is possible to predict which groups are most likely to adopt suicide terrorism not only by understanding religious networks but also through a better understanding of the organizational capital possessed by groups and the relationships between groups. The theory provides greater leverage than existing approaches in trying to determine why actors choose suicide terrorism, and how it matters for international politics.

Required Financial Intensity

The financial intensity of a suicide terror campaign is quite low. As noted above, the often-cited statistic for the "cost" of a suicide bomb, from A to Z, is $150 (Atran 2003). While that figure can vary depending on the cost of the explosive, whether it is a car bomb or not, and other factors, the point is simply that the monetary cost per unit of the hardware for a suicide attack is extremely low (Atran 2003; Hoffman 2003; Israeli Defense Forces spokesperson 2002).[17] This means that the financial barriers to entry should not prevent any group from adopting.

Required Organizational Capital

Alternatively, a high level of organizational capital is required to adopt suicide terrorism. Using Hezbollah, the first mover, as the metric for defining the adoption requirements reveals large organizational challenges for potential adopters of suicide terrorism.[18] With a start year of 1982, Hezbollah turned to suicide terror early in its history, before it had a set operational profile. This

---

[16] The question of whether this is just an issue for Islamic groups is discussed below.

[17] The loss of the life of the bomber is generally not considered a cost in the same way as bomb parts, although there is a clear human cost.

[18] This is not cooperation for a single suicide attack; it refers to a terrorism campaign that includes suicide attacks.

suggests that the optimal organizational age is low. There is not reliable experimentation data or official Hezbollah doctrine to shed definitive light on the critical task focus component of organizational capital. Yet Kramer (1991) suggests that Hezbollah initially conceptualized its mission broadly, which made it open to suggestions, possibly from the Iranians, about suicide attacks. In general, for terrorist groups, unlike nation-states, strong linkages seem to exist between the organizational age and critical task focus components of organizational capital, especially for younger groups.[19] Younger groups, lacking an operational profile due to a lack of attack experience, are likely to also lack a set critical task focus, making them more likely to adopt new innovative tactics.

Even beyond Hezbollah's experience, adoption requires a high level of organizational capital, especially for older groups. Recruiting suicide bombers is a social as much as a physical process; the extreme nature of the act, since it guarantees death for the actor, requires a large investment in training to convince someone to sign on (Iannaccone 2006, 12). The terrorist group has to decide that utilizing suicide terrorism will help accomplish its goals, requiring an evaluation of, among other things, the relative instrumental and/or symbolic benefits, the relative cost of training suicide bombers versus training for other types of terrorist operations, and the potential repercussions in terms of reprisals.

Second, more bureaucratized groups with multiple decision levels and veto points, those with older organizational ages, are likely to have more trouble shifting tactics to adopt. Actors will have political capital invested in particular tactics, especially if their credibility in the group is built on expertise in a specific area, like remotely detonated explosives. Just as members of national militaries often resist the introduction of new technologies or organizational practices that threaten their organizational status by making their training and expertise less relevant, some members of well-established terrorist groups will have strong bureaucratic reasons to resist the introduction of suicide terrorism.[20]

Third, since suicide terrorism by definition involves the death of members of the terrorist group, and potentially members with substantial expertise and knowledge depending on the situation, it cuts into organizational knowledge and expertise (Berman and Laitin 2005). This impact varies depending on whether long-term members or new recruits are used for suicide missions. But in general, suicide terrorism imposes a net organizational cost that has to be balanced out by either the direct instrumental or signaling benefits of the attack.

Finally, there must be people not only willing to die for a specific cause but also willing to kill themselves for it.[21] The suicide terrorist does not want to die

[19] It is also possible that an experienced terrorist group could maintain a broad critical task focus, making it more open to innovation.

[20] Strong top-down leadership could potentially circumvent this problem.

[21] The perception that a "supply" of suicide bombers might not exist could cause a group not to use a tactic. Then again, the decision by a group to use suicide bombing could generate a supply of bombers if the group is popular or does a good job of selling the need for suicide operations.

in the way an individual committing suicide does. Rather, the suicide terrorist has to die to accomplish a mission.[22] This is a supply issue—finding people willing not simply to risk death but to kill themselves in pursuit of an organizational objective as well.

The software costs of suicide terrorism, the costs borne by the organization for suicide bombing, therefore far outstrip the hardware costs. Given the high levels of organizational capital and low levels of financial intensity required to adopt suicide bombing, adoption capacity predicts that groups lacking a high level of organizational capital will be unlikely to adopt.[23]

Incentives to Adopt

The question of incentives to adopt, as opposed to the ability to implement, is different for the suicide terrorism case than for those involving national militaries because terrorist groups face life-and-death struggles on a more daily basis than most national militaries. Most states, most of the time, are not at war and are not mobilizing for it.[24] Terrorist groups, on the other hand, exist on the basis of their violent opposition to the government, and are in a relatively constant state of high vigilance and mobilization. So while for national militaries there is substantial variation in the interest that a military organization is likely to show in a given innovation, terrorist organizations facing the constant threat of extinction should have inherent interests in adopting new tactics like suicide terrorism that may make success against an adversary more probable.[25]

Groups facing asymmetrical military disadvantages in comparison with a nation-state try to find equalizers to at least partially redress the severe military imbalance. Not all groups that utilize suicide terrorism, however, appear to do so because they have literally no other military options. Hezbollah's suicide terrorism campaigns have occurred during times of relative organizational strength, the LTTE in Sri Lanka utilized suicide terrorism simultaneously with a host of other military tactics, and Al Qaeda chose to employ suicide

[22] While there are examples in everyday society, like a bystander pushing a child out of the street to avoid a car and personally taking the blow, of "sacrificial" behavior that goes beyond individual suicide, and most people would likely say that there are values they would die for, actually instrumental suicide terrorism is an infrequent occurrence in the scope of history (Iannaccone 2006, 9–10).

[23] This refers to operations using conventional explosives. It is possible to argue that a suicide attack using a weapon of mass destruction (nuclear, biological, and/or chemical) might be extremely financially costly.

[24] States may always prepare to defend themselves, but that is distinct from mobilization for imminent war.

[25] There also could be a selection effect whereby the groups that adopt suicide terrorism appear to succeed not because it is a useful strategy but rather because since terrorist groups think it is a useful strategy, those with the organizational capabilities to adopt it are also likely to be good at other things as well, meaning they are more likely to succeed for other reasons.

terrorism prior to the U.S. attack on Afghanistan (Gambetta 2005a, 260–61). So while suicide terrorism is certainly a weapon of the "weaker" side in a conflict, it is not always a weapon of last resort. This also proves that it makes sense to think about the organizational adoption of suicide terrorism as a strategic choice rather than an automated response. But even if suicide terrorism is adopted purely out of necessity, the strategic failures of some groups that did not adopt the innovation also suggest that there is utility to looking into the factors that predict adoption.

The question of countering or offsetting strategies for terrorist organizations that do not adopt the innovation is also somewhat distinct from the challenges facing nation-states. Terrorist groups only rarely fight each other, with some exceptions like Hamas versus Fatah at times, meaning terrorist groups rarely have to worry about countering an innovation the way a nation-state has to worry about countering the innovation of another state. As explained above, terrorist groups can coordinate in ways that promote the distribution of information. Yet the relatively constrained resources that face terrorist groups and the localized nature of most terrorist struggles do not, in theory, lend themselves to broad-based international cooperation.

So terrorist groups, for the most part, do not have to worry about countering suicide terrorism and have constrained alliance options. This means that when faced with an innovation, a terrorist group's limited options make its decision process simpler than that of a nation-state. Essentially the only question is whether or not to adopt the innovation.

Suicide Terrorism Diffusion Predictions

Since the financial requirements for adopting suicide terrorism are low but the organizational capital requirements are high, adoption-capacity theory predicts that the innovation should spread widely to those groups with the organizational capacity to adopt, all other things being equal. In particular, given the low financial barriers to entry for suicide terrorism, much of the variance in the adoption or nonadoption of suicide terrorism should be explained by organizational capital. What should this look like in terms of the three indicators of organizational capital: organizational age, critical task focus, and experimentation?

It is difficult to find systematic evidence on research and development or experimentation by terrorist groups. The existing evidence is anecdotal in description and means that the experimentation indicator faces coding constraints. Because nonstate actors face more serious budget constraints than nation-states, and are less likely to have a formal research and development arm, finding formal evidence of experimentation is difficult. Where any evidence of experimentation by terrorist organizations *does* exist, it should correlate with higher organizational capital levels.

The critical task indicator is applicable to terrorist groups. The extent to which terrorist groups view their existence as bound up with specific fighting methods, as opposed to broader goals, influences the breadth of the critical task focus. Those groups with a strong identification to particular ways of fighting, like kidnapping or hijacking, may find it especially difficult to expand their critical task to adopt suicide terrorism. Alternatively, those groups more broadly focused on goal accomplishment rather than methods should have an easier time adopting.

Finally, organizational age is both a powerful explanatory variable and possible to operationalize. As groups build an operational history, they will develop institutionalized command-and-control structures focused on the types of operations that the group conducts. This will naturally make it harder to shift and adopt suicide terrorism, because adoption will challenge established organizational hierarchies. The time gap between the creation of a terrorist group and the first-moving Hezbollah campaign is fortunately measurable.

One other factor that could potentially influence the ability of actors to adopt suicide terrorism is their exposure to direct or indirect information channels regarding suicide terrorism. Where there are direct links between terrorist groups, knowledge and suicide tactics should diffuse faster. As explained above, this is the constrained terrorist group equivalent of an alliance between states that facilitates the spread of an innovation.

Media globalization from 1982 to the present has lowered the informational barriers for groups seeking to mimic the tactics of other groups. Descriptions of suicide bombings, including fairly substantial details about tactics, are available through print media, television, and the Internet. While infiltration methods and specific construction ideas for the creation of suicide vests or car bombs might still remain opaque, much is available in open-source media. The quality of linkages, in terms of the ability to get trusted information and acquire tacit knowledge from more experienced operators, should influence the adoption of suicide terrorism.

Research Design

This chapter examines the diffusion of suicide terrorism through two methods: a statistical analysis of the universe of terrorist organizations from 1968 to the present, and illustrative cases of terrorist group decision making in the wake of the suicide terrorism innovation.[26] The universe of cases is all terrorist groups from 1968 to 2006 as defined by the American Memorial Institute for the

---

[26] A discussion of the selection for the illustrative cases is presented below in the section with the cases.

Prevention of Terrorism and the RAND Corporation (MIPT-RAND) through July 15, 2006, supplemented by illustrative examples of terrorist group decision making in the wake of the suicide attacks innovation.[27] The data set is based on a long-term terrorism data collection effort undertaken by the RAND Corporation and records all types of terrorist incidents, both suicide and nonsuicide.[28]

The aggregated terrorist group information available through the MIPT-RAND data set yields 823 terrorist groups and limited aggregated data on each group, including the group's start date, the motivations of the group, the targets of their attacks, and the total incidents, injuries, and people killed. Only groups that conducted some sort of attacks within the suicide attacks era are included, to avoid biasing the results by including groups that rose, acted, and fell prior to the real debut of the innovation. After the controls added below, the total number of groups drops to 233.

The dependent variable for the statistical analysis is whether or not a group used suicide terrorism. It is coded as a 1 if the group adopted suicide terrorism and as a 0 otherwise. The dependent variable is coded based on data from Pedahzur (2005) and Pape (2005).[29] The main independent variable of interest, a measure for the organizational capital of each terrorist group, is based on its organizational age.[30] Organizational age is defined for these purposes as the time gap between the creation of the terrorist group and the beginning of the suicide terrorism era. Since the suicide terrorism era began with the first campaign, that of Hezbollah, in 1982, each group is coded by its start date in relation to 1982. So the PIRA, since it was formally instituted in 1969, is coded a −13 while Hamas, created in 1988, is coded a 6. The exact form of the age variable does not affect the result.

There may be some particular instances where terrorist groups go through major transformations in response to either internal or external challenges, but defeat is typically not an opportunity for reconstruction in terrorist organizations. A nation-state can often recover from devastating military defeat. Defeat does not always mean a country is conquered and fully integrated into

---

[27] Since the relevant terrorist attacks are not always international and suicide attacks in particular have empirically not always been international, the ITERATE data set, which only codes international incidents, is inappropriate for these purposes.

[28] The selection into the data set is based on the MIPT-RAND definition cited above (Terrorism Knowledge Base, 2006). The question of potential biases in the data is assessed below. Since MIPT only evaluated international terrorist groups prior to 1998, it lacked the entire suicide attack universe. Data from Pape (2005) and Pedahzur (2005), independent from MIPT-RAND, corrected for this limitation.

[29] Assistance from Pedahzur, in particular, was essential in the completion of the coding scheme.

[30] The group start dates are drawn from MIPT data, supplemented with data from Pape.

another state, so it makes sense to reset the organizational age of militaries when major military defeats occur. Nevertheless, terrorist groups in most cases cease to exist once defeated, meaning the organizational age assumption made for coding purposes is accurate. There are potentially a few exceptions, like the PIRA in the late 1970s or Al Qaeda after the invasion of Afghanistan in late 2001, where the level of organizational transformation might be an argument for resetting the group's organizational age. In both cases, though, the leadership remained relatively intact and the goals of the group remained similar, meaning even for those groups the general coding rule makes sense.[31]

Another set of independent variables comes from the MIPT data on group "motivations." The motivations are: anarchist, antiglobalization, communist/socialist, environmentalist, leftist, nationalist/separatist, other, racist, religious, right-wing conservative, and right-wing reactionary (Terrorism Knowledge Base 2006).[32] For each possible motivation, a dummy variable is coded 0 if MIPT did not define the group as having that particular motivation and 1 if the motivation is applicable.[33] Based on the discussion above, I expect religion to serve as a focal point for the transmission of information about the innovation, meaning adoption may cluster around religious networks. To account for potential clustering in particular geographic areas or between particular groups, several control variables are added. To account for the reaction to the U.S. invasion of Iraq, an Iraq War variable is coded 1 if the main country of origin for the group is Iraq and the start year for the group is 2003 or after, and 0 otherwise. To account for the clustering of groups surrounding strife in Lebanon and Israel, a Lebanon variable is coded 1 if the country of origin is Lebanon and 0 otherwise, and an Israel variable is coded 1 if the country of origin is Israel (including Gaza and the West Bank) and 0 otherwise. Finally, many contend that Al Qaeda has played a prominent role in promoting suicide terror tactics among loosely affiliated groups, so an Al Qaeda network variable is coded 1 if Pedahzur, Pape, and others argue that there is a

---

[31] If the coding rule is not useful, the results will demonstrate the limitations of the approach.

[32] After the corrections described below, there are seventy-five religiously motivated groups, of which sixty-nine are Islamic. This means the religion variable is already itself a reasonable proxy for Islam. But I add a specific Islam variable below and describe the results.

[33] According to the MIPT coding scheme, groups can have more than one motivation, which solves the problem of having to decide which motivation is prominent for groups that, for example, are motivated by both religion and nationalism. For the religion variable specifically, the definition is "Religious terrorists commit acts of terrorism in order to comply with a religious mandate or to force others to follow that mandate" (Terrorism Knowledge Base, 2006). This limits the definition to those groups fighting for explicitly religious reasons, rather than including all groups that happen to be religious. For example, though the PIRA was a Catholic group fighting against a Protestant regime, it is not coded as religious, probably because it was not attempting to create a government run on the basis of Catholicism.

TABLE 6.1
Summary Statistics for Terrorism Data Set

	Minimum	Maximum	P50	Mean	Standard deviation
Use of suicide terrorism	0	1	0	0.176	0.382
Organizational age	0	51	22	21.159	14.097
Religion	0	1	0	0.326	0.470
Religion* organizational age	0	51	0	4.052	8.763
Communist/socialist	0	1	0	0.232	0.423
Leftist	0	1	0	0.069	0.253
Nationalist/separatist	0	1	1	0.541	0.499
Other	0	1	0	0.073	0.261
Lebanon	0	1	0	0.073	0.261
Iraq War	0	1	0	0.060	0.238
Israel	0	1	0	0.056	0.230
Al Qaeda link	0	1	0	0.082	0.274

strong link, including shared members and operational planning, between Al Qaeda and a terrorist group, and 0 in other cases.

Since the unit of analysis is the terrorist group and the subject of interest is how groups that conduct *campaigns* behave, the data used for the statistical tests are limited to only those groups that conducted more than one attack and whose attacks have killed at least one person. The conceptual focus of the tests is the decision-making process of terrorist groups, specifically strategic decisions about whether or not to adopt suicide terrorism. Given inherent data collection limitations due to natural secrecy about the organization of terrorist groups, there is a high propensity for red herring groups within the data, meaning groups that never really existed or which only existed on a limited scale. Multiple attacks and fatalities are concrete symbols of group existence that help weed out false groups named for distraction purposes or single-act collaborations labeled as terrorist groups post hoc. If a group has announced its existence but never committed an attack, only engaged in a single operation, and/or never executed a successful attack that generated fatalities, it does not have an operational profile and is excluded from the analysis.

STATISTICAL ANALYSIS

Since the dependent variable is dichotomous, the appropriate statistical model is logistic regression. The models presented below use Huber-White robust standard errors to deal with potential heteroscedasticity. Table 6.2 shows a

TABLE 6.2

Statistical Relationship between Organizational Capital and the Adoption of Suicide Terrorism

	Model 1: Bivariate logit		Model 2: Add group motivations		Model 3: Add geography and linkage controls		Model 4: Interaction	
	Coefficient	Robust standard error	Coefficient	Robust standard error	Coefficient	Robust standard error	Coefficient	Robust standard error
Organizational age	−0.066***	0.016	−0.045**	0.019	−0.048**	0.023	−0.105***	0.036
Religion	—	—	2.298***	0.596	1.887***	0.621	−3.225***	0.915
Organizational age* religion	—	—	—	—	—	—	0.095**	0.044
Communist/socialist	—	—	0.962	0.918	1.194	0.887	0.806	0.869
Leftist	—	—	1.053	1.284	1.336	1.396	0.742	1.486
Nationalist/separatist	—	—	0.468	0.419	0.729	0.537	0.678	0.544
Other	—	—	0.309	1.227	0.267	1.098	0.299	1.103
Lebanon	—	—	—	—	1.574**	0.774	1.595**	0.724
Iraq War	—	—	—	—	0.799	0.675	0.321	0.722
Israel	—	—	—	—	1.943**	0.971	2.009*	1.044
Al Qaeda link	—	—	—	—	1.874**	0.761	1.800***	0.802
Constant	−0.407	0.279	−2.409***	0.63	−3.008***	0.71	−0.441	0.705

Notes: Model 1: N: 233, Wald chi2(1): 17.55, Prob > Chi2: 0, Pseudo R2: 0.108, Log pseudolikelihood: −96.649. Model 2: N: 233, Wald chi2(6): 36.67, Prob > Chi2: 0, Pseudo R2: 0.220, Log pseudolikelihood: −84.525. Model 3: N: 233, Wald chi2(10): 44.87, Prob > Chi2: 0, Pseudo R2: 0.303, Log pseudolikelihood: −75.580. Model 4: N: 233, Wald chi2(11): 50.36, Prob > Chi2: 0, Pseudo R2: 0.325, Log pseudolikelihood: −73.170.

* $p < .10$ ** $p < .05$ *** $p < .01$.

series of statistical models that build from a bivariate analysis of the relationship between organizational age and suicide terrorism to a model that includes all of the independent variables described above, including the group motivation, geographic, and Al Qaeda link variables. While the discussion above underscores that all of the included variables have a theoretical justification for inclusion, the results are presented in building block form to demonstrate the significance of the causal argument across different specifications.[34]

The first three models validate the organizational capital hypothesis, showing that there is a negative and significant relationship between organizational age and the adoption of suicide bombing by terrorist groups. While the coefficient is not especially large, the variable is always significant above the .05 level. The relationship also holds whether or not group motivations, controls for geographic regions, and terrorist group cooperation controls are included or excluded.

As predicted, the adoption of suicide terrorism also seems to cluster generally along the lines of a particular group motivation: religion. Building on other findings showing the potential importance of cultural factors in driving diffusion, religion appears to function as a transmission mechanism for tacit knowledge about suicide bombing. In contrast, nationalism, Pape's key variable of interest, is not significant.[35] Religion can function as a unifying factor for groups to communicate across national or even ethnic lines because it is a medium for the spread of operational knowledge. As a test of the shared religion proposition, I substitute an "Islam" variable for those religiously motivated groups that are Islamic, along with an interaction term between Islam and organizational age. Substituting those variables for the religion and religion*organizational age interaction, I then reran model 4 in table 6.2. As described above, sixty-nine of the seventy-five religiously motivated groups espouse Islamic beliefs. The Islam variable is negative and significant, while the interaction term is positive and significant, just like the religion variable and interaction term in model 4. These results further show how Islam serves as the basis for diffusion between religiously motivated groups. The evidence presented here suggests that suicide tactics have diffused among Islamic groups because that is the coordinating network and probably also because suicide attacks began with Hezbollah. But the evidence does not say anything

---

[34] To ease in the creation of relative risks below, in model 4 the religion variable was transformed to be 0 for religious groups, and 1 for nonreligious ones and the interaction term generated using that specification. This has no effect on the interaction term or its significance, though the coefficient does reverse. The significance level of the lower-order interaction terms is in many ways a function of where the particular specification hits the likelihood function and can change based on respecifying one of the two interaction variables by adding ten, dividing by three, and so forth. The important thing is the significance of the interaction term (Braumoeller 2004; Horowitz, McDermott, and Stam 2005).

[35] This could also indicate processes similar to those described by Elkins and Simmons (2004).

about the propensity for Islamic groups in general to adopt. While about 25 percent of the suicide attack adopters are not Islamic, there are not observations of religiously motivated, non-Islamic groups adopting. Therefore, it is not possible to directly test whether there is something about Islam—as opposed to other religions—that makes diffusion more likely. The results reveal that as hypothesized, the transnational character of religious beliefs—as opposed to nationalism—makes religion an ideal network for diffusion across time and space. Unfortunately, these limits to the data make the results more suggestive than anything else. Figure 6.3 below attempts to shed light on the question of diffusion relationships.

Yet even among religious groups, the high level of organizational capital required for adoption should make it difficult to adopt for groups that cannot learn from a group that already has expertise in organizationally preparing for suicide bombings. This is different from the actual mechanics of suicide attacks. It is the organizational jump to a system that not only encourages group members to actively kill themselves rather than just engage in risky activities, but also has replacement and leadership-retention mechanisms, that requires the greatest effort. One likely example of direct diffusion comes from Hezbollah and Hamas. In 1992, Israel captured and deported 415 Palestinians, mostly members of Hamas but also some members of Islamic Jihad, to Lebanon. While in Lebanon, the Sunni members of Hamas and Shiite members of Hezbollah apparently began coordinating training, leading to the direct diffusion of knowledge concerning suicide attacks from Hezbollah to Hamas (Ricolfi 2005, 91–92).[36] Additionally, confirming Victor Asal and R. Karl Rethemeyer's finding (2008), alliances, in this case between Al Qaeda and other groups, play a significant role in spreading the innovation. None of the other group motivation identifiers are significant.

Given the potential for religiously motivated groups to serve as critical nodes for the diffusion of suicide attacks, whether through direct interaction or indirect demonstration effects, it makes theoretical sense to interact the organizational age and religion variables. The interaction can better differentiate the actual mechanism through which suicide bombing spreads and the significance of the organizational capital variable. Model 4 in the table above shows the results from a statistical model—once again using logit regression—that incorporates an interaction term between the religion and organizational age variables into model 3. The results underscore the importance of evaluating suicide attacks from a diffusion perspective. The interaction between the organizational age and religion variables is significant, and in the predicted direction.

Figure 6.1 displays a graph of the substantive relationship between organizational age and the adoption of suicide terrorism, separated by whether or not

---

[36] A lack of firsthand evidence means deriving conclusions from those interactions is tentative.

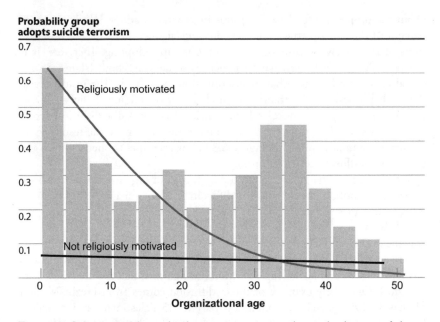

**Probability group
adopts suicide terrorism**

Figure 6.1. Substantive relationship between organizational capital, religion, and the adoption of suicide terrorism.

religion is a motivating factor for a terrorist group. There are two lines on the graph, one for nonreligious groups and one for religious groups. The background displays a histogram of the organizational age variable. The wide variation of values shows that the results make sense across the range of organizational age values. If the distribution was too skewed in one direction or another it would suggest that the sample sizes are too small in the other direction to make valid inferences.[37]

Adoption-capacity theory is unique in its ability not just to explain why groups adopt suicide terrorism—which many other scholars have examined—but also why some groups fail to adopt. For the first five years of its existence, the probability that a religiously motivated terrorist group adopts suicide ter-rorism is between 40 and 60 percent. The probability declines after that point, however, reaching 27 percent by year fifteen and only 15 percent by year twenty-two. In year fifty-one, the oldest age for which data exist in the model, the probability of adoption by religiously motivated groups declines to less than 1 percent. Religious affiliations serve as the diffusion networks through

---

[37] For presentation reasons, the organizational age variable is transformed into a completely positive distribution, running from 0 for groups created in 2006 back through the oldest groups. This has no impact on the results and is only done to improve their presentation.

which knowledge spreads, just as the existing alliance between the United States and Great Britain did in the nuclear weapon case. While terrorist groups do not generally form alliances for "protection" purposes, some do appear to exchange information that leads to the spread of knowledge necessary for adoption. Organizational age appears to explain why groups like the Islamic Liberation Organization in Egypt did not adopt suicide bombing in the mid-1980s and why groups like the Moro Islamic Liberation Front still have not really adopted suicide bombing today.

The way increasing organizational age creates veto points is one explanation for why Fatah, despite its leading role among the Palestinians during the study period, lagged behind younger groups like Hamas and Palestinian Islamic Jihad in adopting suicide bombing. It was only in the Second Intifada with the creation of the Al Aqsa Martyrs Brigade that Fatah adopted suicide terrorism—and this was after a period of "confusion in the organization's ranks" (Pedahzur 2005, 53). One possibility is that the expertise of Fatah at hijacking, assault, and other nonsuicide operations led to a narrowing of the way that Fatah conceptualized its critical task in addition to broadening the number of organizational actors that could function as veto points and prevent adoption.

Nonreligious terrorist groups also become marginally less likely to adopt as their organizational age increases, but religious groups have the most pronounced relationship. For nonreligiously motivated groups, the probability of adoption in the first five years never exceeds 7.5 percent, but still declines over time.

Table 6.3 depicts relative risks and odds ratios derived from model 4 above. The relative risk scores validate the graphic evidence displayed in figure 6.1. Young religiously motivated terrorist groups adopt suicide terrorism in their early operational stages over 60 percent of the time, representing a relative risk increase of almost 640 percent compared with the mean values. At high organizational age levels, the effect reverses; the relative risk of adoption for religiously motivated groups is –71 percent. The large negative relative risk scores for both of the "high" organizational age possibilities show that the organizational age variable does influence the probability of adoption—otherwise the probabilities would be constant across levels of organizational capital and only vary based on religion.

Interestingly, in groups with high organizational ages, indicating terrorist organizations with substantial longevity, the probability of adoption for nonreligious groups is actually slightly higher than for religious groups. The relative risk of adoption by nonreligious groups with high organizational ages is –29 percent, over 40 percent higher than the risk of adoption by religiously motivated groups of a similar age. One explanation is that the availability of tacit knowledge through networks of religiously motivated groups means the nonadopter groups in the data set are like Fatah prior to the Al Aqsa Intifada. These are groups that had extensive operational experience before the onset of

TABLE 6.3

Relative Risks and Odds Ratios Describing the Relationship between
Organizational Capital, Religion, and the Adoption of Suicide Terrorism

Condition	Probability of adopting suicide terrorism	First difference with mean value	Relative risk	Odds ratio
Low organizational age, religiously motivated	0.606	0.524	639.58%	17.238
Low organizational age, not religiously motivated	0.083	0.001	1.01%	1.011
High organizational age, religiously motivated	0.024	−0.058	−70.64%	0.276
High organizational age, not religiously motivated	0.059	−0.023	−28.58%	0.696

Notes: Probabilities generated using Clarify (King, Tomz, and Wittenberg 2000). Probabilities are compared to the mean value. "High" and "low" refer to shifts from the minimum to the maximum values, and vice versa

the suicide terrorism era, rather than those created after the early to mid-1980s that decided not to adopt the innovation. Nonreligiously motivated groups may only acquire the capacity to adopt the innovation later in their existence, since they did not have immediate informational access through religious knowledge networks.

One objection to these results is that the adoption of suicide terrorism is a secular trend; it is not that groups with younger organizational ages have been more likely to adopt the innovation, but rather that only in recent years has the innovation become popular, making it more likely for new groups to adopt. Table 6.4 lists the organizational ages of all adopting groups during their first year of adoption. It shows that even early in the suicide terrorism period, there is a substantial correlation between low organizational age and adoption. It is true that the organizational age of adopters has declined in recent years as more new groups have adopted suicide terrorism—but that is accounted for by adoption-capacity theory, since the current period is arguably the middle bulge in the classic diffusion S curve. The lack of many adopters with high organizational ages at any point lends credence to adoption-capacity theory.

Another objection might be that nearly all of the groups that have adopted suicide bombing are religious, meaning there is not enough variation on that variable to make accurate inferences. Of the fifty terrorist groups that have committed suicide bombings at the group level, forty-one qualify as having both committed more than one attack and generated fatalities during the attacks. Still, of those forty-one groups, almost 25 percent (ten) are coded as

TABLE 6.4
The Correlation between Organizational Age and Suicide Terrorism Adoption

Suicide adoption year	Organizational age at time of adoption															
	0	1	2	3	4	5	6	7	8	9	10	14	17	22	25	44
1982	1															
1983								1								
1985				1												
1987					1											
1989														1		
1993							1									
1994	1															
1995				1									1			
1998										1	1					
1999															1	
2001							1									
2002	1		1						1							1
2003	2		1	1	1											
2004	3	1	1					1	1	1						
2005	5	1	1			1										
2006		1											1			

nonreligious by the MIPT, including Amal (Lebanon), the Kurdistan Workers' Party (Turkey), and the LTTE (Sri Lanka).

Additionally, while the logit results demonstrate a significant relationship, both statistically and substantively, the appropriateness of the logit form is assumed rather than derived from the data. It is possible to object that the curve for religiously motivated groups in figure 6.1 reflects just a few younger terrorist adopters that are skewing the overall relationship between organizational age, religion, and the adoption of suicide terrorism. To test for this possibility, I ran a form of nonparametric regression known as Lowess regression on the set of religiously motivated groups.[38] Lowess regression involves generating point-by-point estimates based on the information near each point, instead of imposing a global functional-form assumption on the statistical relationship. The outcome, displayed in figure 6.2, shows a good fit between the original curve and the Lowess curve. This increases confidence in the results.[39]

[38] The regression is run for religiously motivated groups instead of nonreligiously motivated groups, since that is where most of the variation happens so it is where the largest possibility for error exists.

[39] While the curve is somewhat less steep in some places, the general pattern is nearly identical. For more on Lowess regression and the particulars of the specification in figure 6.2, please contact the author at horom (at) sas.upenn.edu.

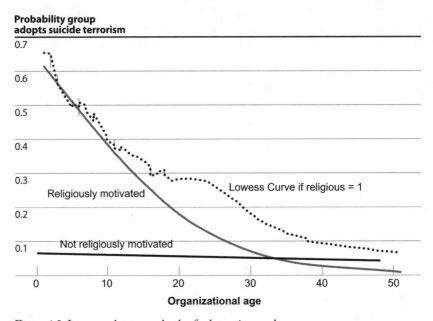

Figure 6.2. Lowess robustness check of substantive results.

A final objection to these findings could be that older terrorist groups like ETA are those that have had political wings, meaning they had a great deal invested in the political process and sought to avoid disruption through suicide attacks. But a first mover, Hezbollah, and another prominent adopter, Hamas, both have political wings, and conducted suicide attacks after the creation of those wings. Also, even after groups build inroads into the political system, if they are still employing terrorist tactics in any form, it is important to ask why they would choose to not use a method generally considered effective. A terrorist group, for example, might not adopt because they want to limit the moral outrage and number of casualties from their attacks, and just use attacks to signal continuing capabilities and help push along a bargaining process. Nevertheless, it is just as likely that the failure to adopt might be interpreted as weakness with limited casualties from attacks signaling an inability to deliver punishment, making it harder to achieve group goals because the government will feel less pressure to negotiate. So even if a group has a political wing, it should not make it significantly less likely to adopt suicide terrorism.

Figure 6.3 highlights the relevance of applying a diffusion approach to the study of suicide terrorism. It includes most of the known groups that have conducted suicide terror operations, grouping some in the same regions, and

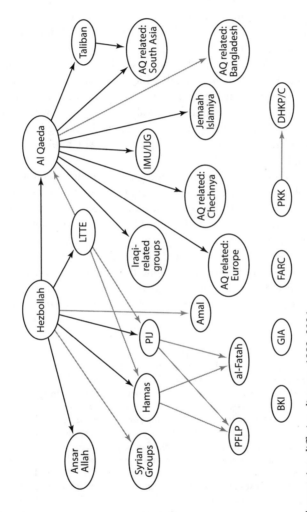

Figure 6.3. Suicide terrorism diffusion diagram, 1982–2006.

attempts to describe the relationships between the groups. The figure depicts two types of relationships, with the direction of the arrows in the figure symbolizing the direction of the relationships. The black arrows signify a direct relationship, meaning there is evidence of meetings, joint training, and other behavior that would indicate the potential for the direct diffusion of suicide terror knowledge from one group to another. The gray arrows represent an indirect relationship, meaning there is evidence that a group learned, through print media or otherwise, about suicide terror tactics from information about the behavior of another group.[40]

Suicide terrorism has diffused from two main "hubs" over the course of the era. The first hub is Hezbollah, through which the Palestinian organizations and the LTTE adopted. The second comes from Al Qaeda, which learned from Hezbollah but then became a central node through which multiple jihadi groups around the world appear to have learned. Having links to one of these hubs seems to play a major role in predicting which groups will adopt suicide terrorism.

The linkages show the importance of recognizing the intersections between groups; it is not a coincidence that nearly every single group that has used suicide terrorism over the last twenty-five years has a direct or indirect link to other suicide terrorist groups, and often more than one. One could argue that the linkages merely reflect groups going out and seeking knowledge after they have made a strategic decision to adopt. The fact that groups need to seek assistance in so many cases to effectively adopt, though, proves that there is tacit knowledge associated with the innovation that is relevant for adoption. The point is not that diffusion-related forces are the sole determinants of adoption, just that they matter.

---

[40] Appendix 1 documents the links between groups and explains all of the abbreviations. Figure 6.3 also raises the question of connections between nonsuicide adopters. To answer that question, I gathered data from the MIPT on the connections between all groups. Combined with the data on the adoption of suicide attacks, I created three variables for each group, a number of connections variable, a number of connections to suicide attack adopters variable, and a ratio of the two. The raw connections data reveal that on average, adopters have significantly more ties to other groups than nonadopters (verified through a t-test). More important, the ratio data show the percentage of connections that a given group has to groups that have adopted suicide attacks. A score of .5 means 50 percent of a group's connections are to groups that have adopted suicide attacks, while a score of .1 means that only 10 percent of a group's connections are to groups that have adopted. The mean score for suicide adopters is .43, compared to .19 for nonadopters, and that difference is significant according to a t-test. This indicates that there are significant differences in the connection patterns of suicide groups versus nonsuicide ones. While this analysis does not cover the indirect connections shown in figure 6.3, it suggests, based on the available evidence, that figure 6.3 is not simply an artifact of the data on suicide attack adopters. Yet there is an endogeneity problem in the data, since they do not show the timing of connections between groups, so it is not possible to tell whether groups connected before or after they adopted suicide attacks. While these data are potentially useful, care has to be taken in interpretation.

Specific Cases

Each of the following cases includes an outline of the group and its activities, an analysis of its organizational capital level and links to groups with suicide terrorism experience, and then an assessment of how the introduction or non-introduction of suicide terrorism influenced the fortunes of the group. It is important to consider groups across the adoption spectrum, from fast adopters to nonadopters. Otherwise the factors that suicide terrorist groups have in common might not really explain the existence of suicide terrorism, since they might be factors also held in common with groups that did not use suicide terrorism. The cases below include the LTTE, a non-Muslim group that adopted suicide terrorism; the PIRA, a prominent nonadopter of suicide tactics; and Al Qaeda, the most infamous adopter of suicide terror tactics.

The LTTE

The group that conducted the single-largest number of suicide attacks between 1981 and 2003 was the LTTE. It conducted 191 attacks from 1981 to 2003, or almost as many as every group in the Middle East combined (224).[41] The LTTE emerged in the late 1970s and early 1980s as a radical militarized splinter group from the more moderate Tamil United Front (Pedahzur 2005, 80; Ramasubramanian 2004; Tekwani 2006). The Tamils are an ethnic minority mostly centered in northern Sri Lanka, in the Jaffna region, that feels oppressed by the Sinhalese majority.

The LTTE began substantial guerrilla operations against the Sri Lankan government in 1983, mostly targeting military and governmental outposts. At the time, the LTTE may have had direct interaction with Hezbollah. It trained with the PLO in Lebanon in the late 1970s and early 1980s, and likely directly observed Hezbollah during its fight with Israel. Velupillai Prabhakaran, the LTTE's leader, explicitly viewed Hezbollah's successful attack on the U.S. Marine barracks in 1983 as a model for leveraging the Sri Lankan government (Hoffman and McCormick 2004, 259; Hopgood 2005, 51; Narayan Swamy 1994, 97–101).

The first suicide terrorism operation by the LTTE, a car bombing by Captain Millar on July 5, 1987, probably took place in response to Operation Liberation, a sustained assault by the Sri Lankan Army (SLA) into Tamil territory in May 1987. Unlike other groups that use suicide bombs as an independent means of attack, Millar's truck bomb was designed as the first wave of a broader LTTE assault—the bomb exploded in an SLA camp in Nelliady, and was followed by an LTTE assault on SLA forces that disrupted Operation Liberation (Hopgood 2005, 50).

[41] This information is drawn from a database on suicide attacks created by Ricolfi (2005, 82).

The political timing and the method of the first LTTE suicide attack are consistent with the generally instrumental nature of LTTE suicide bombings. The vast majority of suicide attacks by the Black Tigers and Sea Tigers, the LTTE land and sea suicide attack squads, were focused on "hard targets," bases and people for which other strategies would not succeed, like the 1991 assassination of Rajiv Gandhi in Tamil Nadu (Pedahzur 2005, 77).

ORGANIZATIONAL CAPITAL

How does the LTTE fare on the organizational capital scale? While the Tamil struggle predated the beginning of the modern suicide terrorism era, the LTTE insurgency against the Sri Lankan government started in late 1982 and early 1983, at the onset of the suicide terrorism era. So the organizational age variable predicts that the relatively young LTTE would have the capacity to adopt. The LTTE also defined its critical task broadly. Its violent genesis at the beginning of the suicide terrorism era, at a point when both more established groups like the PIRA and newer groups like Hezbollah were all active, may have given the LTTE an especially broad perspective on guerrilla warfare strategies. This made it a strong candidate to adopt.

Two other factors also potentially disposed the LTTE to adopt suicide terrorism. First, while support from the Tamil population was necessary for the LTTE to get recruits to the cause, the LTTE did not require public support in the same way as groups such as Hamas. The LTTE carved out its own geographic space in the Jaffna region, complete with its own military bases. Unlike the PIRA, which depended on soft support from the local population to conceal its hideouts and provide emergency safe havens, the LTTE did not need local popular support to maintain its existence. The Sri Lankan military was physically unable to eradicate the LTTE, meaning that simply knowing the location of military bases was not enough to destroy the group. Additionally, since LTTE suicide attacks were generally aimed specifically at military and governmental targets, they did not place the local population intentionally at risk (Hopgood 2005, 59). While LTTE suicide attacks did result in civilian casualties, focusing on noncivilian targets probably gave the LTTE much greater leeway in the public sphere in dealing with the consequences.

Second, the LTTE command structure was centrally oriented with Prabhakaran, who held an intense level of operational control. The LTTE command structure appears to have been set up with direct control from higher levels but flat hierarchies below the top, enabling the adoption of suicide terrorism without substantial "middle management" or intermediate-level veto points.

EXTERNAL LINKS

As explained above, Tamil fighters spent time in the Middle East in the late 1970s and early 1980s training in the Bekka Valley with the PLO and probably Hezbollah. Prabhakaran also reportedly indirectly learned from abroad, using the Marine barracks attack and its aftermath, when Western forces withdrew from Lebanon, as an example of the type of coercive campaign that was most likely to lead to success for the Tigers (Hoffman and McCormick 2004, 259). These direct and indirect links between the LTTE and Hezbollah may well have aided in the LTTE's grasp of operational concepts surrounding suicide terrorism, since it was familiar with the operational landscape facing groups in the Middle East. Tactical innovations by the LTTE improving the effectiveness of suicide bombing operations then diffused back from the LTTE to the Middle East. For instance, in response to governmental efforts to detect suicide bombers, the LTTE developed the suicide vest (Jackson et al. 2007, 68). This microinnovation soon spread to Middle Eastern groups like Hamas and Palestinian Islamic Jihad.

ASSESSMENT

The use of suicide attacks by the LTTE served a vital role in its efforts to defeat governmental forces and achieve its goals. In particular, the LTTE's terrorist activities played a direct part in forcing the Sri Lankan government to negotiate in 2002 (Jackson et al. 2007, 61). Brian Jackson and his colleagues cite a former Sri Lankan foreign service officer as saying, "This is an example where terrorism has succeeded. We have been cowed. We have been intimidated by suicide terrorism. It is that simple. The fear caused by this tactic has made us cave into them" (66). So while the use of suicide terrorism does not always guarantee success, adoption by all accounts played a significant role in boosting the fortunes of the LTTE during the 1990s and early years of the twenty-first century.[42] After peace talks collapsed in 2006, however, violence restarted and culminated in the defeat of the LTTE by the government in 2009.

Al Qaeda

Al Qaeda is less a group than a network of affiliated believers in over forty-five countries who have greater or lesser degrees of direct ties (Pedahzur 2005, 97–98).[43] It is generally considered a terrorist organization with an "unusually innovative nature" (Byman 2003, 142). This section focuses specifically on the Osama Bin Laden–led segment of Al Qaeda that existed in Afghanistan through the 2001 U.S.-led invasion, and is likely still centered in the border area between Afghanistan and Pakistan.

---

[42] It is obviously somewhat impossible to answer the counterfactual question of the success of the alternative tactics that the LTTE might have utilized if it had not used suicide terrorism. Given that even Sri Lankan officials pointed to suicide terrorism as an effective tool for the LTTE, however, the balance of evidence seems to overcome the counterfactual possibilities.

[43] I am indebted to Stephanie Kaplan for making this point especially clear.

After the Soviet evacuation from Afghanistan in 1989, a generation of mujahideen trained to fight against the Soviet occupation turned its attention elsewhere. The United States itself, rather than the corrupt and apostate Arab regimes it supported, became a target following the deployment of U.S. military forces on Saudi soil in the run-up to the first Gulf War in fall 1990 (Benjamin and Simon 2002, 105–7).

In the mid-1990s Osama Bin Laden (1996) issued a fatwa declaring a jihad, or holy war, against the United States in response to its "occupation" of land on the Arabian Peninsula. In 1998, he issued another statement titled "Jihad against Jews and Crusaders" (World Islamic Front 1998). Moving from the Sudan to Afghanistan, Bin Laden received safe haven from the Islamic Taliban regime. In the pre-9/11 period, before the U.S.-led military actions forced massive organizational changes, Al Qaeda efficiently deployed its military assets. As Stephen Holmes (2005, 157) writes, "Their basic rationality can be inferred from their decision to mobilize fast-moving light forces acting in secrecy. This is a highly efficient way to deploy the modest resources at their disposal."

Al Qaeda followed a two-prong operational strategy in the pre-9/11 era. In the early 1990s the organization became involved in guerrilla conflicts around the globe. From Bosnia to Chechnya to the Philippines, sending money and training fighters for export abroad, Bin Laden attempted to steer Muslims everywhere on to a more radical pathway guided by jihadist ideals (Benjamin and Simon 2002, 127–28, 145–47). On the other hand, in the late 1990s Al Qaeda itself began conducting operations, focusing on large-scale operations like the embassy bombings in Africa, the bombing of the USS *Cole*, and the September 11, 2001, terrorist attacks in the United States. Those operations initiated and directed by Al Qaeda, as opposed to just linked to or partially financed by Al Qaeda elements, like the Khobar Towers and first World Trade Center bombings, have exclusively been suicide terror attacks (Brassey's 2004; Benjamin and Simon 2002; Bergen 2002; Gunaratna 2002; National Commission on Terrorist Attacks upon the United States 2004).[44]

ORGANIZATIONAL CAPITAL

Al Qaeda is a textbook case of a group with organizational attributes that made adoption likely. According to Daniel Benjamin and Steven Simon (2002, 102–3), Bin Laden started forming the core leadership of the group that became Al Qaeda following the departure of Soviet troops from Afghanistan in 1989. This places the creation of Al Qaeda after the onset of the suicide terrorism age, meaning natural obstacles that might exist for an older terrorist group did not exist for Al Qaeda. Prior experiences in Afghanistan, Egypt, and other locations did not bias it toward particular operational tactics, which might

---

[44] Numerous government reports including that of the 9/11 Commission address the motivations of Al Qaeda, which is why this section focuses more on the diffusion of suicide terrorism to Al Qaeda.

have militated against the adoption of suicide terrorism, for two reasons. First, the Afghanistan campaign was a guerrilla warfare campaign directly for territory, making it qualitatively different from Al Qaeda's indirect terrorist efforts. Its tactical experiences did not lead to constricting operational biases. Al Qaeda leaders understood that their "new" campaign was different and would require a distinct approach. Second, the need for asymmetrical tactics against conventionally stronger Soviet forces guided the strategy of the mujahideen. The lesson learned from Afghanistan was the ability of a smaller and allegedly weaker group of fighters to defeat a stronger opponent if they took advantage of all the possible options. This prevented the institutionalization of a narrowly defined critical task focus and made the adoption of a weapon maximizing "will," which Al Qaeda had in abundance, more likely.

Al Qaeda's organizational design also made it more difficult for veto points to form and block change. Al Qaeda is infamously based on a network structure with high levels of operational autonomy, although it shifted from a more hierarchical model to a flatter one after the September 11 attacks and U.S. response (McAllister 2004). This deliberate approach prevented the generation of bureaucratic layers that might block tactical transformation toward suicide terrorism. The "virtuality" of Al Qaeda, its use of information age tools to communicate and plan, gave it the ability to maintain a presence in multiple locations, continually scout targets, and plan operations without centralizing too much of the group in one place (Benjamin and Simon 2002, 169; Byman 2003, 155).[45] In the pre-9/11 era, Al Qaeda maintained multiple, overlapping network structures. While some involved high levels of central control, others entailed ties as loose as the short-term training of operatives who then returned to homelands across the world.[46]

Al Qaeda defined its critical task from the beginning quite broadly: it wished to bring about the defeat of the West and the return of an Islamic caliphate. Daniel Byman (2003, 147) maintains that "al-Qaeda regards attacking as a goal in and of itself, demonstrating the movement's strength and determination." The view of all attacks as ends rather than just means, whether in terms of the signaling, instrumental destructive, and/or theological values, made Al Qaeda tactically flexible; there was no reason to exclude new and innovative tactics on the grounds that they were unproven.[47] A statement from Ayman

---

[45] It is important not to overstate the extent to which the networked nature of Al Qaeda mattered. Networks can be more or less hierarchical, and involve more or fewer interlinkages (McAllister 2004, 307).

[46] Hoffman identifies four Al Qaeda operational levels: the core cadre, trained amateurs, walk-ins, and financing/inspiration for theologically/ideologically similar groups in other locations (Hoffman 2002; McAllister 2004, 304).

[47] The theological commitment to radical jihadism also dovetailed with justifications for suicide terrorism by emphasizing the importance of jihad as an individual commitment and martyrdom through violence (Benjamin and Simon 2002).

al-Zawahiri, Al Qaeda's second in command, demonstrates the high degree of commitment to new methods, and highlights the link between Al Qaeda's cost-benefit analysis and suicide terrorism:

> The mujahid Islamic movement must escalate its methods of strikes and tools of resisting the enemies. . . . In this regard, we concentrate on the following: . . . the need to concentrate on the method of martyrdom operations as the most successful way of inflicting damage against the opponent and the least costly to the Mujahedin in terms of casualties. (cited in Sageman 2004, 23)

EXTERNAL LINKS

Al Qaeda had both direct and indirect links to suicide terrorist groups, meaning that it had clear exposure to suicide terrorism tactics. In Sudan in the 1990s, Bin Laden met with Hezbollah leader Imad Mughniya to discuss operational concepts, and Al Qaeda operatives "visited Hezbollah training camps in southern Lebanon" (Benjamin and Simon 2002, 127). The 9/11 Commission's final report, quoting multiple captured Al Qaeda operatives and declassified U.S. intelligence documents, states that Bin Laden sent Al Qaeda operatives to Hezbollah specifically for the purpose of learning how to execute an attack like the U.S. Marine barracks bombing in Lebanon (National Commission on Terrorist Attacks upon the United States 2004, 68, 470–71). This shows the direct diffusion of knowledge from Hezbollah to Al Qaeda.[48] Al Qaeda operatives returned to the Sudan with training manuals and tapes from Hezbollah on bombing operations. The U.S. embassy bombings in Kenya and Tanzania in 1998 utilized large car bombs and tactics similar to those perfected by Hezbollah in the 1980s (Pedahzur 2005, 100–101).

The USS *Cole* bombing in October 2000 likely demonstrated indirect operational learning from the Tamil Tigers. The Sea Tigers pioneered sea-based suicide attacks using explosive boats, just like the USS *Cole* attack (Gunaratna 2002, 74; Pedahzur 2005, 100).[49]

ASSESSMENT

Al Qaeda's high level of organizational capital made it a strong candidate to adopt suicide terrorism. In addition to the culture of martyrdom that Bin Laden cultivated among Al Qaeda members based on his interpretation of Islam, Al Qaeda possessed a high degree of organizational flexibility. Its suicide attacks, like those of Hezbollah and the LTTE, clearly played a significant

---

[48] Even if Bin Laden had already thought about the embassy bombings as suicide attacks, he had to export the training of operatives for the mission to achieve his goal.

[49] Of course, Al Qaeda has also provided financial and material support to a litany of groups abroad from Chechnya to the Philippines, but that has involved Al Qaeda being the source of suicide terrorism diffusion, rather than the object.

role in bolstering the instrumental effect of its major terror attacks. The 9/11 attacks might not have occurred absent suicide terrorism. It is difficult to see how else Al Qaeda could have generated the impact necessary to take down the Twin Towers and knock down part of a Pentagon wall with regular explosives; the 1993 World Trade Center attack, a nonsuicide truck bomb, actually proved that only a suicide attack would succeed.

The PIRA

The failure of the PIRA to adopt suicide terrorism is something of a puzzle. If the essence of Pape's theory (2005) about democratic occupiers with different religious beliefs becoming prime targets for suicide terrorists is true, the Catholic PIRA should have turned to suicide terrorism against the Protestant United Kingdom. Similarly, Bloom's theory (2005) about outbidding and market share might predict that the PIRA would turn to suicide terrorism to compete with the "Official" IRA on the Left and the "Real" IRA on the Right, especially during fragile periods of the peace process. But the PIRA never adopted suicide terrorism.

Beginning in 1969, the "Provisional" IRA (PIRA), stating that it was the successor to the "authentic" IRA Army Council, the government of Ireland, split from the Official IRA and initiated a violent terror campaign against targets in northern Ireland, the United Kingdom, and occasionally further abroad, including Washington, DC, and Zaire (Bell 1990; Coogan 1993; Jackson 2005, 94). The PIRA primarily targeted British military and government installations and figures, along with some civilian areas in Belfast and England, using three primary methods of attack: explosives, grenades launchers, and surface-to-air missiles (Jackson 2005, 98). The campaign only wound down in 1997 following the cease-fire that led to the Belfast (Good Friday) Agreement.[50]

Despite a legacy of violence over thirty years that began before the suicide bombing era and continued for more than a decade after the Lebanon campaigns, the PIRA never utilized suicide terrorism. In three cases in October 1990, the PIRA did use kidnapping to try to blackmail people into committing suicide attacks, but the nonvoluntary nature of these acts explains why none of the major data sets of suicide terrorism—the MIPT, Pape, or Pedahzur—code these three cases as acts of suicide terrorism. The PIRA has never directly engaged in suicide bombings.[51] Since the PIRA's operations continued

---

[50] In 2005 the PIRA formally renounced the use of armed force and encouraged all of its members to abandon armed struggle.

[51] In the first case, the PIRA kidnapped the family of Patsy Gillespie and forced him to drive a car armed with a remotely detonated bomb into an army checkpoint. The explosion killed Gillespie and five police officers. In the second case, the PIRA threatened to kill relatives of James McEvoy unless he drove a car bomb into an army checkpoint. While he drove the vehicle as requested, he survived the explosion. In the third case, the PIRA purportedly held the wife and child of a man

throughout the 1980s and 1990s, it had ample opportunity to observe suicide terrorism operations by other groups throughout the period, so what explains its failure to adopt this tactic?[52]

ORGANIZATIONAL CAPITAL

One reason why the PIRA failed to utilize suicide terrorism, though not the only one, likely had to do with the mismatch between the organizational capital required to adopt and the organizational capital possessed by the PIRA. Most important, the PIRA's operational success criteria, or the way that it evaluated when and how to conduct operations, probably made it difficult for the PIRA to adopt. In other areas, the PIRA did demonstrate operational creativity, including its perfection of the car bomb. Still, the mismatch between organizational design and the adoption requirements for suicide terrorism made adoption difficult. Based on official PIRA documents, Jackson (2005, 112–13) identifies the operational success criteria for the PIRA as: "volunteer safety," "security force casualties," "economic damage," "publicity and public reaction," and the "minimization of civilian casualties."[53]

PIRA operational procedures placed a premium on preserving the expertise and dedication of trained volunteers to the exclusion of riskier operations. Given that cost-benefit analysis for nonsuicide attacks often led to the abandonment of high-risk operations that might place PIRA volunteers at risk of capture and death, it is not surprising that the PIRA did not adopt a tactic whose fulfillment necessitated the death of group members. Also, given the indiscriminate nature of most suicide terrorism, the explicit limitation on civilian casualties represented a substantial limitation in the way the PIRA defined its mission, making adoption less likely.

The PIRA's high organizational age and command structure also made the adoption of suicide terrorism less probable. The PIRA's campaign began in 1969, twelve years before the Lebanon bombings. Its belief that it was the legitimate heir to the IRA Army Council, the military arm of Ireland, led the PIRA to adopt organizational procedures similar to those of a regular military, even describing its units as brigades and battalions (Jackson 2005, 96). The PIRA had a command staff, and was separated into operational and nonoperational subcomponents (logistics, intelligence, etc.). Four organizational layers—the General Army Convention, Army Executive, Army Council, and General Headquarters—stood above the active unit commands and nonoperational

---

hostage unless he drove a car bomb into a checkpoint, but the bomb did not detonate (English 2003, 266; Kalyvas and Sánchez-Cuenca 2005, 211).

[52] There is some evidence that Michael McKevitt, the head of the Real IRA, a PIRA splinter group, wanted to create a squad of suicide bombers to attack the British army but could not get recruits (Ayers 2006, 17; Carton 2005).

[53] Volunteer was a term used by the PIRA to describe its members.

commands. This structure undoubtedly made agile operational planning difficult (96).

In the late 1970s, facing increasing pressure from the government and decreasing support from the local population, the PIRA reorganized, shifting from a more military-style structure to a more cell-based or social network one. According to Edgar O'Ballance (1981, 226), prior to the PIRA's reorganization, it extensively studied prominent international terrorist groups and their organizational structures, including the PLO, the Baader-Meinhof Gang, and ETA. For high-profile operations or especially difficult ones such as the assassination attempts on Margaret Thatcher and the Earl of Mountbatten, or the attack on 10 Downing Street, though, General Headquarters appointed and directly controlled special teams of operatives (Bell 1990, 109; Bell 1998, 451; Collins and McGovern 1997, 82–83; Jackson 2005, 115–16).

Thus, both the organizational age of the IRA and its critical task focus made the adoption of a technique such as suicide terrorism less likely. While the IRA engaged in research and training, existing operational concepts constrained the scope of those efforts. It engaged in "sustaining" tactical innovation, not strategic innovation (Jackson 2005). The PIRA excelled at adopting incremental or even substantial innovations in familiar areas of operations that involved getting better at producing the types of attacks it already produced. Nevertheless, when it came to producing different types of attacks through a new method—that is, suicide terrorism—the organization did not adjust.

EXTERNAL LINKS

In the mid-1970s, the PIRA also had a collaborative relationship with other terrorist groups. Sean O'Callaghan (1998, 196) argues that "the links between ETA and the [PIRA] run deep; the two organizations have often cooperated and pooled ideas, technology and training." The PIRA also trained with other terrorist groups in the Middle East and Libya, developing especially strong ties with the PLO and the Libyan government (Jackson 2005, 120). Yet two factors cut against these ties leading to a diffusion of suicide bombing techniques. First, the most extensive cooperation between the PIRA and groups that adopted suicide terrorism techniques was in the 1970s, before the suicide bombing era began. Second, PIRA operatives have downplayed the extent to which cooperation with other groups, especially the Libyans, influenced their operational art. One PIRA member, talking about cooperation, said, "Don't mind that talk about Libya. The Libyans were trained in conventional warfare. They couldn't teach us anything" (cited in Coogan 1993, 363; Jackson 2005, 120).[54]

---

[54] Of course, PIRA volunteers might purposefully understate the importance of cooperation with terrorist groups abroad to make themselves look stronger and more independent. But they might also have incentives to overplay aid from abroad to make it look like they have allies and are therefore stronger. Given this uncertainty, it makes sense to take the statement at face value.

Assessment

Stathis Kalyvas and Ignacio Sánchez-Cuenca (2005, 219–200) contend that the dependency of the PIRA on core constituencies that opposed civilian casualties, the Catholic population of northern Ireland and other supportive Irish Catholics abroad, forced the PIRA away from more destructive attack possibilities that placed civilians at risk, including suicide terrorism.[55] But the public support argument is an incomplete explanation of the PIRA's failure to adopt. Even if most suicide terror operations today in places like Israel are targeted at civilians, the LTTE and the early Lebanese suicide terrorists, which almost exclusively targeted government and military installations, show that suicide bombings do not necessarily involve maximizing civilian casualties. So the assertion that a fear of alienating the public due to the collateral damage from suicide attacks deterred the PIRA is incomplete.[56]

Another alternative argument is that the PIRA was not interested in suicide terrorist attacks. It possible that the PIRA had decided on a strategy of negotiation as the suicide terror era was dawning, meaning it was not a candidate for adoption. The hunger strike in 1981 led by Bobby Sands, when seven PIRA (and three Irish National Liberation Army) members deliberately and publicly starved themselves to death in prison to protest the British refusal to grant them political prisoner status, however, proves that the PIRA was highly resolved and attempted to signal that resolve to a wider audience through suicidal actions.[57] A lack of PIRA interest is partially consistent with adoption-capacity theory because the high PIRA organizational age and lack of critical task focus could have meant that PIRA members did not fully conceptualize the potential benefits of suicide terrorism correctly.

It is difficult to assess how the PIRA's failure to utilize suicide terrorism influenced the outcome of the northern Ireland conflict. Continued progress in the peace process, which culminated with the disarmament agreement in 2005, potentially may not have occurred if the PIRA had utilized destructive suicide terrorist attacks that further inflamed public opinion against it and led to an escalation of operations. Then again, suicide terrorist attacks might have brought the United Kingdom to the bargaining table sooner and more seriously, causing

[55] Irish Catholic outcry at civilian deaths like in the Harrods of London bombing in 1983, according to the Kalyvas and Sánchez-Cuenca story (2005, 220), caused restraint that prevented the PIRA from using suicide terrorism.

[56] One could argue that "Western values" held by the target audience in the United Kingdom and Ireland would have spurred revulsion that made the achievement of PIRA goals less likely. But such an explanation would contradict the rationalist understanding of behavior promoted by most scholars who would make the public opinion argument in the first place.

[57] Also if the evidence shows a lack of interest by the PIRA in suicide terrorism, rather than a lack of ability to implement the tactic, it potentially hurts the Kalyvas and Sánchez-Cuenca theory by showing that it was not a lack of core constituency support that prevented the PIRA from implementing suicide terrorism.

a more rapid peace on better terms. The counterfactual argument is plausible in either direction. What is clear is that the PIRA's lack of organizational capital handicapped the ability of the PIRA to adopt disruptive lessons from abroad, making it ill equipped to adopt suicide tactics.

CONCLUSION

Adoption-capacity theory can help explain the development and spread of suicide bombing, showing that the theory has applications beyond major powers and even nation-states. Financial and organizational constraints similar to those that influence the decision making of military organizations also influence the decision making of terrorist groups. Second, evaluating the adoption of suicide terrorism from an innovation perspective helps capture the distinctive characteristics of suicide terror adopters in comparison with nonadopters. While previous research has accurately described many of the factors that motivate people and groups to adopt suicide terrorism, a more complete theory that can explain why some groups adopt suicide terrorism while others do not has been lacking. The interaction with religion highlights the importance of network effects in suicide bombing. The extensive interlinkages between religiously motivated suicide terror groups and the demonstration effects that have fueled adoption by nonreligiously motivated groups show that focusing only on internal politics or the strategic environment falls short. Adoption-capacity theory fills the gap, revealing how the high organizational capital requirements for adopting suicide terrorism made those terrorist groups that were most successful in the presuicide terror era unlikely to adopt the new innovation.

Kalyvas and Sánchez-Cuenca (2005) explain terrorist group strategy as a function of its need for and level of public support, arguing that a U-shaped curve best depicts the nonincidence of suicide terrorism.[58] For groups that are totally disconnected from the population, like Al Qaeda, suicide terrorism imposes low organizational costs because the organization does not need local support to survive. For groups interlinked with the population, such as Hamas, societal culture shifts in ways that make suicide terrorism more acceptable. Yet for groups in the middle, like the PIRA, that are partially linked to the population but more operationally distinct than a group such as Hamas, the risk of civilian casualties inherent in suicide bombings makes them too costly. The group fears angering the local population and losing the support that is necessary to shield it from authorities, meaning it will not adopt suicide bombing (Kalyvas and Sánchez-Cuenca 2005).

---

[58] Kalyvas and Sánchez-Cuenca do not lay out an explicit definition of terrorism as either violence by nonstate actors or violence that targets civilians. Since they group activities by insurgents and terrorist groups together, however, they likely adopt a nonstate-based perspective.

The Kalyvas and Sánchez-Cuenca theory is not inconsistent with adoption-capacity theory. Adoption-capacity theory does not rule out that popular support could influence the interest of terrorist groups in adopting suicide terrorism, only that organizational constraints will prevent many groups from adopting and predispose others to adopt. Still, an explanation for strategic choices that relies entirely on popular support ignores the internal organizational factors that influence terrorist groups. So while their theory and adoption-capacity theory make similar predictions for several groups, adoption-capacity theory more fully describes the decisions of more groups. For example, Kalyvas and Sánchez-Cuenca cannot accurately explain the Tamil Tigers, the group that engaged in the single-largest number of suicide attacks in the 1981–2003 period. As the evidence above shows, the Tamil Tigers were about as linked to the local population as the IRA or even less so—and a lot less than Hamas and a lot more than Al Qaeda. So according to the Kalyvas and Sánchez-Cuenca theory, the LTTE should not have engaged in suicide terrorism.

The Kalyvas and Sánchez-Cuenca contention is also limited by its focus on terrorist groups as individual actors in a vacuum, rather than as linked actors in the international system. Given the evidence of terrorist group cooperation and knowledge distribution, even between groups of different religions and with different goals, it makes sense to view adoption as influenced by a diffusion process, instead of as an entirely independent decision made by each terrorist group.

Another alternative explanation might claim that prominent groups like the PLO, the PIRA, and ETA have failed to adopt because they are actually pro-tostates attempting to eventually operate within the regular international system.[59] The statehood theory is similar to the Kalyvas and Sánchez-Cuenca theory, which states that groups that need public support but whose publics do not support suicide bombing will be less likely to adopt. Like their argument, the statehood one presumes that the need to look "legitimate" to the public and international community causes restraint. As with the Kalyvas and Sánchez-Cuenca notion, I think the statehood assertion has some merits in particular cases. While it can explain something like the nonadoption of the independence-seeking PIRA, though, it cannot explain the adoption of a group like the Tamil Tigers in Sri Lanka, which also had aspirations to statehood and adopted. Additionally, Hamas, like Fatah, has aspirations to statehood and adopted even though Fatah did not initially adopt. Thus, while the question of statehood aspirations is a good one, it does not provide a complete explanation for the diffusion (and nondiffusion) of suicide terrorism.

Finally, some might regard suicide terrorism as a special case of insurgency warfare, where actors facing overwhelming force choose asymmetrical responses

---

[59] Thanks to an anonymous reviewer for making this point clear.

because they are most likely to be effective. Ivan Arreguin-Toft (2005) maintains that groups are functionalist and choose the strategies most likely to be effective. If this theory is true, the adoption of suicide terrorism by terrorist organizations should vary solely with perceived success by groups that use suicide terrorism, not based on organizational factors. Adoption-capacity theory does not exclude the possibility that perceptions of success influence adoption; indeed, that is a key part of the contention. But according to a purely functionalist argument, the PIRA and ETA should have adopted the innovation. Both had a desire to inflict casualties through terrorist attacks, and the LTTE clearly showed that suicide terrorism could be effectively utilized without targeting civilians. Yet neither adopted—an outcome only fully explained by adoption-capacity theory due to their high organizational ages and the bureaucratic constraints that influenced their behavior.

Chapter 7

CONCLUSION

In 1452, Sultan Mehmed II, the ruler of the rising Ottoman Empire, had a problem.[1] While the Ottoman Sultanate had conquered most of the former Byzantine Empire's territory, the walls of Constantinople, the crown jewel of Asia Minor, still held. Mehmed II worried that despite the enormous naval and land forces he had amassed, including Bashi-bazouks and Janissaries, he could not breach the walls and conquer the city. His advisers brought a Hungarian engineer named Urban before him; Urban purported to have the ability to build the largest gun ever constructed for the purposes of battering down the walls of Constantinople. While the Ottoman military already had cannons, its production capabilities were reportedly inferior to the best in Europe. Urban told the Ottoman sultan, "Am I able to cast a cannon capable of throwing a ball or stone of sufficient size to batter the walls of Constantinople? I am not ignorant of their strength; but were they more solid than those of Babylon, I could oppose an engine of superior power: the position and management of that engine must be left to your engineers" (cited in Gibbon 1788/1974, chapter 68, 755). Given funding and authority by Mehmed II, Urban took charge of the Ottoman foundry at Adrianople and began construction of his cannon. On its completion, the cannon was brought by Ottoman engineers to the edge of the Bosphorus and, according to legend, it sank a Venetian ship with one shot. Mehmed II was impressed and authorized the construction of a still-larger cannon. Workers completed the cannon in time for the Siege of Constantinople, which began on April 2, 1453. Legend says the gun, named Mahometta, weighed almost a thousand pounds, and took between 60 and 140 oxen to move, 100 people to aim, and 2 hours to reload (Cipolla 1965, 94). The siege succeeded; the city fell on May 29, 1453.

There are two morals to this story relevant to this book, and neither of them is whether or not the cannon actually worked (it did not: on the second day of the siege the gun cracked, and by the fifth day at the latest it apparently no longer functioned). First, the Ottomans gained access to this particular cannon production technology through the diffusion of engineering knowledge from a Hungarian engineer. Mehmed II attempted to overcome a potential strategic weakness by adopting a new innovative technology that, combined with his siege tactics, could help him achieve his goals. Second, despite the impressive

---

[1] Mehmed II is also spelled Mehmet II.

cannon, it was the combination of Ottoman military planning, with its waves of attacks that still left the disciplined and highly trained Janissaries in reserve, and advances in indigenous Ottoman military technology, that determined the outcome of the battle, not the quick-fix technological option.

This story, told again and again in military history, demonstrates the importance of both technological and organizational factors in producing new advances in military power. The combination of new organizational approaches to the employment of force and new technologies can produce increases in military power that are much larger than the sum of their parts. The quantitative and qualitative tests presented in the preceding chapters show that adoption capacity, the combination of financial intensity and organizational capital possessed by a state, influences the way states respond to major military innovations and how those responses affect the international security environment.

It is also important to recognize the limitations of this book. This book does not cover the entire universe of major military innovations, and the theory does not address every factor that motivates state behavior. Nevertheless, the consistent pattern of evidence across the cases suggests that failing to account for the diffusion of military power distorts the overall picture of international relations. Within the broader context of thinking about how states make strategic choices under conditions of uncertainty, this book attempts to explain issues ranging from the rate and scope of diffusion for particular military innovations to the circumstances in which shifts in relative power are most likely to occur and escalate to war. This conclusion explores the implications of adoption-capacity theory for international relations theory, and then applies it to debates in the United States and abroad about the future of warfare.

THE CONTRIBUTION OF ADOPTION-CAPACITY THEORY

Adoption-capacity theory posits that variations in the financial and organizational requirements for adopting an innovation govern both the system-level distribution of responses and the way individual actors make decisions, as well as the subsequent implications for international politics. Financial intensity refers to the level of resource mobilization required to adopt a major military innovation. Organizational capital represents the intangible capacity of organizations to adapt to a changing strategic environment. While measurement issues limit their application in some cases, these concepts can help identify why different types of military innovations spread differently. Analyzing system- and state-level behaviors allows the theory to make predictions about changes in warfare, and the balance of power in general, along with the way particular innovations will affect different types of states. This builds on previous research on the spread of military power by drawing on elements of diverse arguments to build a more cohesive understanding of both diffusion and its consequences.

Financial and organizational requirements drive the rate and extent of diffusion while making alternative strategic choices like forming alliances more attractive or less attractive. For example, the railroads/rifles/telegraph innovation introduced by the Prussian military in the mid- to late nineteenth century had a low unit cost and there was preexisting commercial interest in subcomponents like the railroad. The organizational requirements, the delegation of more tactical control to officers on the battlefield, proved something of a challenge, but not overwhelming. Within ten to fifteen years of the Franco-Prussian War in 1871, most European militaries with an interest in adopting the innovation had adopted or had the capacity to do so (Herrera and Mahnken 2003). In contrast, the uniquely high financial and organizational requirements associated with carrier warfare have formed a large-scale barrier to entry even for some of the most powerful navies in the world.

Previous macrolevel explanations about leaders, politics, and power that do not take into account the adoption requirements of the dominant forms of military power at the time and how they influence the strategic choices of key states are incomplete. But this is not to say that geostrategic requirements do not matter; far from it. The geopolitical interests of a state likely play a large role in shaping its decision about how to respond to an innovation. But understanding the external strategic environment facing a state is not enough to predict its behavior in response to a major military innovation. Explanations focused on whether a state or leader feels threatened often only predict whether or not a state will respond to an innovation, not the content of that response. States can bandwagon with the first mover, balance against the first mover with another likely adopter, reduce their international profile by moving toward neutrality, attempt to counter the innovation in ways besides adoption through their own military means, offsetting its relevance, and/or try to adopt only a portion of the innovation and use it in a different way. For example, in the case of nuclear weapons, both Japan and Germany rely on their alliance with the United States and the international nuclear nonproliferation regime instead of building their own nuclear weapons. They decided that the potential benefits of adoption are less than those of an alternative strategy.

Applying the insights of business innovation scholars like Christensen (1997) and Henderson (1993) reveals how financial and organizational requirements for adopting innovations crucially influence who wins and loses in international politics. Innovations requiring higher levels of financial intensity are likely to benefit preexisting powers. The rich get richer when what is most required to generate a new form of military power is cash to buy weapons. Alternatively, innovations with high organizational capital requirements are more likely to disrupt relative power balances, since preexisting powers are the states most likely to have ossified military bureaucracies precisely due to their expertise at a previous method of warfare. Those states will find it difficult to transform since organizational experience will nearly always militate in favor of the status quo. In these cases, first movers often experience magnified advantages. Innovations requiring low levels of financial intensity or organizational capital to adopt might, in contrast, provide late-mover advantages where later adopters can take advantage of the mistakes of earlier adopters and more efficiently integrate military innovations into their national defense architectures.

Explaining the Behavior of Both State and Nonstate Actors

One advantage of adoption-capacity theory is that it can predict not only the behavior of great powers but also that of smaller powers and nonstate actors. Financial and organizational requirements also affect the strategic decision making of nonstate actors such as insurgent and terrorist groups. For example, chapter 6 shows that the high organizational capital requirements for adopting suicide terrorism significantly influence the probability that groups will adopt. One factor, though certainly not the only one, that blocked experienced terrorist groups like the PIRA and ETA from adopting suicide terrorism was their high number of bureaucratic veto points due to the prestige given to group experts in other tactical areas. This made the adoption of a new tactic, suicide terrorism, more difficult. In contrast, groups without preexisting operational biases, like Hamas and Al Qaeda, adopted and achieved successes in part due to their adoption.

Strategic Success and Failure

Adoption-capacity theory also highlights the crucial distinction between when states attempt to respond to a major military innovation and whether or not that choice is likely to succeed. Major military innovations are distinct from simpler changes because they are systems for applying military force, not just individual technologies. So while the old aphorism that "where there is a will, there is a way" may apply to purchasing a particular type of fighter aircraft, this book shows that it is much less likely to apply to something as complicated as carrier warfare. This is especially true when talking about adoption immediately after the introduction of an innovation. Sometimes, no matter the intentions of a state, it will simply lack the capability to adopt a new innovation; financial shortfalls or organizational barriers may prevent adoption even if the state does not realize this at first.

For instance, the Soviet Union attempted to become a first-rate naval power in the midst of the cold war. It had massive financial resources and naval thinkers like Admiral Gorshkov, who had learned from the experiences of Germany, another European continental power that attempted to develop a significant navy. But the Soviet Navy was unable to mount a serious challenge to U.S. surface naval superiority or even produce a high-quality fleet carrier. Part of the problem lay in the mismatch between the organizational requirements for effectively operating the dominant naval weapon of the post–World War II era, the aircraft carrier, and the organizational capabilities of the Soviet Navy. Carrier warfare requires a uniquely high level of organizational capital, given the systems integration challenge of simultaneously operating the boats and an air force at a high training tempo. The weak commitment to experimentation in the Soviet Navy, due to the fear of political consequences for making a mistake, as well as the layers of veto points built into the decision-making hierarchy by the need for political "obedience," left the Soviet Navy ill equipped to build or operate high-quality aircraft carriers.

Due in part to these constraints, for much of the cold war the Soviet Navy pursued a dominant strategy of attempting to counter carrier warfare with advanced antiship missiles, submarines, and land-based aviation. When it deviated from its countering strategy, it failed for the same reasons that it was forced into that strategy in the first place: financial constraints and its lack of adequate organizational capital.[2]

More Efficiently Explaining the Pace and Outcome of Power Transitions

One weakness of many explanations for the rise and decline of states is that they gloss over the specific mechanisms by which change occurs. Adoption-capacity theory helps describe both why power transitions occur and how. When organizationally disruptive innovations debut, those that gain significant advantages in international politics are the first movers that have generated new ways of producing military power. The key here is the demonstration of the innovation, not just the first use of the incorporated technologies. Their actions trigger strategic responses on the part of the rest of the international system, with the resulting balance of power determined by the match between the financial and organizational requirements for adoption and the financial and organizational capabilities of key states, modified by their alliance postures and other options. This gets at the microlevel reasons why a state that seems like a great power may end up lacking the ability to compete in the new military power realm—it lacks the organizational capital to adopt the new innovation—as well as how a rising power ends up overtaking the military capacity of a preexisting power. Clearly this is not the only way that a power transition can happen; power transitions can occur absent a significant military innovation. Nevertheless, it is in periods of uncertainty about the basis of military power that rapid shifts become more likely. One example comes from early modern Europe, where a series of innovations ranging from sieges with cannon and the increasing importance of firepower to the bastion fortress upset the European balance of power, and helped usher in two centuries of more frequent and lethal warfare across the continent.

---

[2] This also shows that states do not always make optimal decisions when figuring out how to respond to innovations.

Explaining the Microfoundations of Changes in Military Power

Detailing the *mechanisms* by which large-scale military change occurs makes adoption-capacity theory a potentially useful microlevel explanation under the rubrics of multiple theories of international relations. Whether it is lawlessness and the need for states to secure themselves against enemies that provides the strongest incentive for responding to innovations, or preferences derived from interest groups in the domestic political sphere, the match between the financial and organizational requirements of an innovation and the capabilities of the state will help to determine response strategies. That is, adoption-capacity theory is agnostic with regard to metatheory because its microfoundations, based on organizational processes and financing, are applicable whether one believes the fundamental driving force of international politics is disorder and external threats or domestic politics. In either case, the organizational and financial capabilities required to adopt the dominant military innovation of a period fundamentally alter which states are likely to succeed in international politics.

For example, in a "liberalism" world (Moravcsik 1997), a large mismatch between the adoption requirements for an innovation and the adoption capacity of a given state might make the arguments of domestic military and civilian interest groups in favor of adoption less successful. Even if proadoption groups succeed in orienting government policy toward adoption, adoption-capacity theory suggests that the institutional conditions would make actual adoption unlikely. In any case, the means by which domestic preferences are translated into national strategy in the military sphere are influenced by the match or mismatch between requirements and capabilities, making some domestic arguments more persuasive than others and affecting which groups coalesce to lobby for particular policies.

Alternatively, diffusion processes are often underspecified in realist interpretations of international politics. In contrast to the prediction that extreme threats will lead to successful emulation, many states facing security threats have failed to adopt military methods even after successful demonstrations. To return to the blitzkrieg example, French military advisers had a firsthand view of German combined arms warfare in the invasion of Poland, yet this knowledge did not cause a substantial change in French military strategy in the months before the German attack on France. Similarly, as Goldman (2006) points out, the decades-long military decline of the Austro-Hungarian Empire and the arguably centuries-long military decline of the Ottoman Empire did not prompt the imitation of crucial military innovations by either empire, despite clear necessity. Both states ceased to exist in their prior forms at least in part due to their failure to adopt key military innovations prior to World War I. Essentially, strategic competition alone can, at best, explain whether or not states will be interested in responding to a new innovation, based on whether they face a threat. Another theory must capture what happens next. Adoption-capacity theory can fill this void, depicting the substance of the response and the implications for international politics.

The Future of Warfare

As described in the introduction, intense debates in the U.S. defense community and around the world continue about the most likely character of future warfare along with the implications for the international security environment. After several years of war in Afghanistan and Iraq, and hard-fought lessons learned about counterinsurgency warfare in both of those locations, one critical debate in the U.S. defense community is between those who view the future of warfare as most likely to be characterized by complicated irregular military operations, and those who want the U.S. military to focus more on conventional military operations and not get too caught up in transforming for counterinsurgency (Gentile 2008; Mazarr 2008; Nagl 2009).

While terms such as military-technical revolutions, revolutions in military affairs, network-centric warfare, and transformation have fallen out of favor in the Pentagon, an underlying question still critical to all interested in the future of war is how the era now known as the information age will influence the creation and use of military power (Krepinevich 2002).

In a rare appearance before Congress in 1995, Andrew Marshall, the influential director of the Pentagon's long-range in-house "think tank," the Office of Net Assessment, argued that two innovations in warfare had the potential to turn into military revolutions. First, deep-strike weapons, precision-guided munitions, GPS surveillance, and other electronics innovations could cause a gradual shift in warfare away from direct ground combat of the type seen in the nineteenth and twentieth centuries, and toward more stand-off, long-distance warfare. Second, information technology could open entirely new nonkinetic areas of warfare (Marshall 1995).[3]

These broad potential areas of warfare may apply differently to state and nonstate actors.[4] One prominent attempt to theoretically link advances in information technology with modern insurgencies is fourth-generation warfare (4GW). Most clearly proposed by Thomas X. Hammes, the premise of 4GW is that there are generations of warfare that feature dominant effective strategies and that each generation since the Napoleonic era has decentralized power to smaller numbers of individuals. In the fourth generation, information technology has

---

[3] The possible innovations discussed in the chapter are those that have received the greatest amount of press. There are a variety of other potential forms that future innovations could take.

[4] The terms nonstate warfare, irregular warfare, and guerrilla warfare are used interchangeably in this chapter to mean warfare that is not between the formal armies of nation-states.

made decentralized coordination and disruption efforts by insurgent/terrorist groups more plausible than ever. These conditions, in combination with the advantages of will and interest often possessed by insurgent groups, have turbocharged the ability of nonstate actors to effectively attack state targets and evade state-led attacks (Hammes 2004, 2–3, 209–11). Many have challenged 4GW as flawed in its description of modern insurgencies as somehow new or different, pointing to the behavior of the Minutemen in resisting British forces in the early stages of the American Revolution as just one example of guerrilla practices prior to the information age (Echevarria 2005; Luttwak 2005; Van Creveld 2005; Wirtz 2005).

A major critic of the idea that the information age may matter much at all is Biddle (2007b, 463–64), who contends that the modern system of warfare, the use of cover, concealment, decentralized troop movements, and combined arms warfare in response to firepower advances in the early twentieth century, has been the "orthodox" method for winning land campaigns for territory since World War I, and that recent fads will not dislodge its importance. But it is unclear what Biddle means when he talks about the modern system as the orthodox form of war. Does he mean modern system armies are most likely to succeed? Or that the modern system is the most prevalent form of organization for battle by nation-states? What about forms of warfare used by nonstate actors? If major powers employ the modern system but other states do not, is it still orthodox? Biddle and others also argue that scholars should not put too fine a point on the "newness" of insurgencies. After all, as an empirical matter insurgencies function best when attempting to drive occupying powers out of the insurgents' home territory (Freedman 2005, 260–62).

Adoption-capacity theory's recognition of the importance of the information age is therefore simultaneously orthogonal and critical to debates about the future of war. While it is foolish to proclaim with confidence exactly how the information age will shape future warfare, it is equally naive to claim it is unlikely to matter at all. The technologies associated with the information age will certainly create change, even though the content of that change is unknown. The debate about whether future wars will be conventional land wars fought with the modern system, stand-off, over-the-horizon, precision strike wars, or insurgency wars sometimes obfuscates the way that the information age is likely to produce shifts in the financial and organizational requirements for warfare in any potential future world. It also ignores the "hybrid" aspects of warfare.

Many influential thinkers now believe that the future of war will be hybrid, with both high-intensity and irregular characteristics (Hoffman 2007). While historical memory tends to define wars like World War II as conventional and ones like Vietnam as irregular, nearly all wars contain multiple segments fought at different levels of intensity and with different optimal strategies (Hoffman 2007). Adoption-capacity theory can help explain which types of actors are likely to benefit, which are likely to flounder, and the possible overall consequences for power balances and warfare, regardless of the specific vision for the future of warfare.

Figure 7.1 breaks down the debate on the future of war within the U.S. defense community on the basis of three questions: whether or not one thinks the information age will transform the character of war in some way, whether or not one thinks the most probable future conflict scenarios are conventional or irregular, and whether one thinks the U.S. military is likely to excel at the type of warfare suggested by the answers to the first two questions. Question one is the broadest because it delineates the range of the possible for warfare by outlining the relevance of potentially new information age tools. Question two follows—once you decide on the range of the possible, what matters is the type of warfare, conventional or irregular, in which the range of the possible most often occurs. Question three focuses specifically on the United States and its ability to succeed given the way someone answers the prior questions.

It is certainly true that defense debates vary from country to country. Different countries may have different security environments, meaning the most likely type of warfare may differ from country to country. Figure 7.1 shows the way that specific disagreements in the assumptions made about the future lead to different visions for force structure, doctrine, and strategy in the United States.[5]

Predictions Derived from Adoption-Capacity Theory for Nation-states

While the evidence is still limited and any predictions are necessarily tentative, there is enough available proof to guess the rough ranges of financial intensity and organizational capital required for some possible innovations of the future. In reality, the impact of the information revolution on warfare is still in its infancy. At this point, for example, information age components have helped some segments of the U.S. military to perform the same tasks they did in the past, only much faster and more accurately. Information technology has generally been employed in a sustaining rather than a disruptive fashion. It has not yet led to large-scale organizational changes or major shifts in thinking about the situations in which force deployments are possible. This section lays out several potential areas of development and uses adoption-capacity theory to explain some of the ways these developments could shape the international security environment. While necessarily speculative and incomplete, this demonstrates the importance of continuing to think about the financial and organizational requirements for generating military power even in a time of uncertainty about the future of war.

---

[5] Many scholars informed this table. In particular, see Biddle 2004; Cohen 1996; Hammes 2004; Hoffman 2007; Kagan and O'Hanlon 2007; Marshall 1995; Nagl 2005; Pillsbury 2001.

Importance of information age for warfare	Most likely future war type	Confidence in American capability	
		Optimistic	Pessimistic
Important	"Regular"	Information superiority, network-centric warfare	Information attacks/ assassin's mace
	"Irregular"	Aghan model: CIA/SOCOM/USAF	4th generation warfare – Hammes/Lind
Not important	"Regular"	Stephen Biddle – modern system	Multi-theater warfare/big army – Frederick Kagan
	"Irregular"	Small Wars Manual – H.R. McMaster	Old-school army Vietnam realists

Figure 7.1. U.S. strategic beliefs about future warfare. The "Afghan model" refers to the war model followed during the initial invasion of Afghanistan in 2001.

PRECISION WARFARE

Precision warfare is an interesting case for measuring financial intensity because of the striking cost differentials between the platforms that deliver advanced munitions, weapons like "smart" bombs and UAVs and the supporting equipment to make information age military strikes possible. As described in the carrier warfare chapter, there are huge expenses associated with building strike forces from capital-intensive platforms. The F-22 would have cost approximately $257 million per plane once its production run is complete, and more since its production run was cut short, according to the Selection Acquisition Report published by the U.S. Department of Defense (Bolkcom 2005, 4).[6] Each B-2 bomber purportedly cost $1.2 billion to build. The actual smart components that help the U.S. military deliver munitions within feet from hundreds of miles away cost much less, however. For instance, one estimate describes the cost of a joint direct attack munition tail kit that adds GPS guidance capabilities to an existing bomb as only $21,000 per unit (U.S. Air Force 2006c). The cost per unit of the new GBU-39 small diameter bomb, the newest smart weapon developed by the U.S. Air Force (2006b), is reportedly only $40,000 per unit. In the future, commercial GPS satellites could allow a country with less advanced technology to acquire at least part of the smart targeting

[6] In his analysis of the FY2007 defense budget request, Steven Kosiak (2006) at the Center for Strategic and Budgetary Assessments writes that the F-22 acquisition cost would have been about $61 billion for 179 planes, or a unit cost of about $340 million per plane. To compare, the purchase of 462 F-18 E/F planes by the U.S. Navy will cost about $44 billion, or about $95 million per plane, and the F-35 is estimated to cost a little over $100 million per plane. So while the F-22 is clearly the most expensive plane in the U.S. arsenal, even so-called cheaper options are still incredibly expensive, especially given that the yearly defense budget of a country like Italy, which has the tenth-largest defense budget in the world, was $30 billion in 2005, (in purchasing power parity 2003 U.S. dollars) according to the Stockholm International Peace Research Institute (Stålenheim, Sköns, Perdomo, and Perlo-Freeman 2006).

capacity possessed by the U.S. military by arming its older planes and ships
with information age equipment (Kagan 2006, 390).[7] For now, though, the
platform cost is still extraordinarily high.

Taking full advantage of precision warfare, presuming a linear technological development trajectory, which history shows is unlikely to hold, would
probably require substantial financial investments in new air, land, and sea
platforms, while also upgrading targeting equipment to benefit from commercial technological breakthroughs. Thus, while partial adoption of precision
warfare will be financially possible for many states, the financial intensity required for full adoption will be quite high.

A turning point in the financial intensity required for adoption may come if
the sophistication of the munitions reaches a level where the platform becomes
less relevant. When the range and guidance capabilities of a cruise missile or
other munitions are sophisticated enough that states can launch them from a
commercial cargo plane rather than a B-2 bomber or strategic missile submarine, it will lower the financial barriers and open the door for a whole new set
of states to rapidly adopt precision warfare.

Predicting the level of organizational capital required for a future innovation is difficult because, as cases like blitzkrieg and carrier warfare demonstrate,
it is generally much easier to figure out trends in the types of technologies
likely to appear in future wars than to figure out the probable uses for those
technologies. Yet it is possible, based on existing research, to make general predictions. The organizational capital required to adopt precision warfare as
practiced by the U.S. military is not especially high; at least for the foreseeable
future it will involve the use of existing or upcoming weapons platforms and
organizational structures. Operationally, incremental precision warfare still involves training for the same types of actions; militaries will just be able to carry
them out more efficiently or adapt the methods slightly to allow for a more
aggressive strategy (for example, carrying out air strikes on a wider range of
facilities simultaneously).

Over time, precision warfare could also become organizationally disruptive
rather than sustaining. If, as discussed above, advanced launch platforms like
bombers and submarines are no longer required to deliver advanced munitions,
meaning cargo planes and ships can substitute for a fraction of the cost, it
could signify the need for a new organizational model. Implementing such an
innovation could require transforming the procurement process away from expensive platforms as well as shifting education and training, because training
pilots, navigators, and submariners, among others, would require a different

---

[7] Additionally, a state like China could probably launch a bare-bones GPS constellation into
orbit much more cheaply than the United States could, if the United States needed to rapidly deploy more satellites. Once sophisticated technologies are employed, it is difficult to dial them back.

skill set. Advanced precision warfare and robotics could also alter force structures as militaries attempt to acquire larger numbers of cheaper platforms and munitions instead of focusing their resources on a relatively smaller number of assets.

The relatively high level of financial intensity but low level of organizational capital required to implement precision warfare in the short-term means that it is likely to sustain and deepen existing international power balances. Like other costly innovations that are organizationally sustaining rather than disruptive, U.S.-style precision warfare developments are likely to benefit militaries with large budgets that have preexisting expertise in air, land, and naval combat. It will be difficult for states besides rising great powers to adopt. Continual technological improvements mean that investments could quickly become obsolete—a risk exacerbated during periods of economic turmoil that strain defense budgets. Alternatively, if precision warfare advances to a level at which the financially intensive platforms previously necessary for operations are no longer required, the result for international politics would be profound. While the major powers, through their financial advantages, would have enough resources to stay ahead in most cases, the nature of their relative advantage would shift. Instead of a qualitative and quantitative edge, in which countries like the United States have both larger quantities of most types of weapons and more advanced weapons than other countries, major powers would merely possess a quantitative edge.

A next-generation precision warfare innovation also could give countries such as China and India an opportunity to catch up to the U.S. military in the area of conventional warfare. If cheaper alternative delivery vehicles can accurately deliver precision munitions, the expertise built over the last several decades in the U.S. defense industry could become much less relevant. This could give indigenous defense industries in rising powers with roaring economies a clean slate from which to militarily compete with the United States.

CYBERWARFARE

Shifting to another potential area of information age warfare, cyberwarfare may require a low level of financial intensity to adopt. Changes in warfare such as the construction of cyberwarfare units could require a fairly low level of financial intensity to adopt due to their heavy linkages to commercial enterprises. While the idea that anyone with a computer and a modem could attack the Pentagon is probably a bit overstated, the increasing production of high-end information technology components abroad, especially in the developing world, will speed the global diffusion of dual-use Internet technologies.

Information technology in general is more like the railroad than the rifle in that there are large commercial incentives for investment independent of military needs (Herrera 2006).[8] The actual cost for a military organization seeking to adopt information technology for military purposes will be relatively low. While the more powerful the computer, the easier some tasks like hacking are likely to be, even relatively low-power commercially available computers can serve as hubs to launch cyberattacks (Arquilla 2003, 219; Arquilla and Ronfeldt 2000, 2).

Cyberwarfare, representing a major innovation if it shifts from being a sustaining backbone to a theater of military operations, will also probably require large levels of organizational transformation to implement in that case. Despite existing research into cyberwarfare and the development of some nascent capabilities, it has not been systematically exploited. Cyberspace is currently a conveyor belt through which information travels to allow war fighters to do their jobs. The shift to treating the cyber realm as a central theater of war would be enormous—potentially similar to the shift from the air being a peripheral surveillance area to a theater of war itself.

The existence of computer security operations throughout the U.S. government, but especially in the U.S. military, where the Department of Defense has repelled thousands of denial-of-service attacks in cyberspace daily since the 1990s, show that the United Sates already takes the possibility of cyberwarfare seriously (General Accounting Office 1996). Yet cyberwarfare currently represents a limited area of investment for the U.S. military and nearly all militaries around the world. Most countries think about cyberoperations in terms of the defensive: defeating attacks by hackers and others attempting to disrupt state-based military operations.

One example of potential developments in cyberwarfare comes from Chinese military analysts, some of whom believe that, due to the great distances required to project power across the Pacific Ocean, the U.S. military is overly dependent on information superiority. Efforts to disrupt U.S. information superiority could therefore help give the Chinese military an edge in a future conflict (Mulvenon 2006, 87–90). While James Mulvenon (91) concludes that existing Chinese analyses of the potential for cyberwarfare underestimate U.S. defensive capabilities and overestimate the potential of Chinese information operations, the Chinese military has begun actively planning for offensive cyberoperations—a nascent recognition of a potential future major military innovation.

The low financial intensity and high organizational capital required to adopt cyberwarfare as a new theater would also present a disruptive challenge. The global diffusion of computing technology means that cost will not be a barrier to entry for most states or even many nonstate actors.[9] As a completely new

---

[8] Of course, military organizations generally supported rail companies in their attempts to lay track because it helped to quickly and efficiently transport troops from place to place.

[9] I explain this point in greater detail below.

area of warfare rather than a modification of a prior area, the relative organizational challenges will be large but not overwhelming. Rising militaries concerned with challenging existing power hierarchies, like China and India, should be especially interested in adopting. Moreover, if the United States is not the first mover, meaning promotion pathways and military educational structures have not changed in advance to pave the way, America's prior lead in more traditional forms of conventional warfare could become a bureaucratic barrier and delay U.S. adoption of more extensive cyberwarfare capabilities. Thus, adoption-capacity theory predicts the steady spread of cyberwarfare throughout the international system. It is even possible that cyberwarfare could upset traditional power balances because the high level of organizational capital required for true systemwide adoption may disadvantage existing powers. Nevertheless, for the most extreme impacts to occur, countries will have to find ways to generate significant kinetic consequences from cyberattacks. While the low level of financial intensity means even states that fail to organizationally transform will be able to purchase the necessary hardware, it will be difficult for them to fully take advantage of the innovation.

ROBOTICS

A third important area, which overlaps with the other two in some areas, though not others, is the application of robotics to war. The clearest example at present is the use of UAVs such as the MQ-1 Predator and Global Hawk by the United States in the war on terrorism. Other current applications of robotics include small robots that can disable improvised explosive devices along the road in places such as Iraq. Though the United States is the global leader in the use of robotics in warfare, over 250 corporations and governments from 35 countries have joined Unmanned Vehicle Systems International, an international nongovernmental organization for groups interested in UAVs (van Blyenbrugh 2008).

The more widespread use of unmanned systems represents a potentially disruptive development in warfare. Focusing just on UAVs, since they have received the most attention thus far, their further development could significantly change the way that nations utilize air power. For instance, high-end estimates for the MQ-1 Predator drone place the production costs at about $6 million per plane (a system including four planes costs around $30 million), compared to over two hundred million dollars per F-22 (Bolkcom 2005; U.S. Air Force 2008).

Several countries already produce UAVs designed for both military and commercial purposes. Commercially oriented UAVs conduct border surveillance, do crop spraying on farms, or engage in other tasks. For a variety of reasons, including a relatively low cost per unit, maneuverability, fuel, and lower training costs, UAVs are likely to increase in importance throughout the armed services (Boot 2006, 440–41; Ehrhard 2001; Haulman 2003). As they become increasingly capable of doing the jobs currently done by manned aircraft, this could cause changes in recruiting, education, and training that reverberate throughout the services. Essentially, the person best qualified to fly a manned fighter plane or bomber may not be the person best qualified to operate a joystick hundreds of miles away from the action. The educational background desired and training regimen required could change dramatically, meaning that the full integration of UAVs could become a disruptive core competency. (Also see Singer 2009.)

For example, while UAV operators in the U.S. Army "consider themselves pilots," they have struggled for acceptance as pilots by their peers (Steele 2006, 26). Disputes over whether UAV operators would get traditional pilot "wings" or medals for successful missions in Afghanistan and Iraq demonstrate how fully integrating UAVs into the U.S. military will create bureaucratic challenges (Kosiak 2004; Office of the Secretary of Defense 2005). It might also matter whether UAVs become replacements for fighter planes in a future major warfare contingency or remain merely useful in close air support operations for small patrol teams tracking terrorist activity (Benjamin 2008). From a global perspective, the widespread adoption of UAVs could change the air power equation. The lower unit costs mean many more countries could produce larger air forces, and countries such as Japan and South Korea with strong robotics industries could potentially become leaders in the production of UAVs. In the most extreme case, the strategic environment could shift from one in which only a few states can produce fighters and bombers to an environment in which many states can produce systems that are close to each other in quality. This would shift the relative advantage of the United States to one in which it tries to stay just ahead of other states and has to maintain larger numbers of planes rather than relying on the significantly greater quality of each plane.

Nonstate Actor Predictions

While this section has mostly focused on the consequences of the information age for state warfare, the information age has already affected how states and nonstate actors interact. The future of warfare may involve much more frequent low-intensity struggles between national military forces and nonstate actors than warfare between states. If the lesson of the first Gulf War, according to Indian Army chief of staff General Krishnaswamy Sundarji, was not to go to war with the United States without nuclear weapons (Schneider 1994, 227), the lesson many might draw from America's struggles at times during the wars in Afghanistan and Iraq is probably that countries should not engage in conventional warfare with the United States at all, but that other means of challenging the United States are available.

Information-based advances have been far from irrelevant for Western forces conducting counterinsurgency operations in Iraq and Afghanistan. UAVs are used to provide real-time intelligence to small unit patrols and even engage in strikes on individual targets (Benjamin 2008). Real-time communications have improved the situational awareness of U.S. units in the field, though patrolling a city block involves different sorts of information challenges than targeting a group of enemy tanks. Regardless of what one thinks of network-centric warfare in general, U.S. information systems developed from that vision have been useful to commanders in Afghanistan and Iraq (Kagan 2006, 354). Tracking and scanning technologies to protect U.S. ports and airports from weapons smuggled from abroad are another way that the information age may matter for states in situations beyond state-on-state warfare. Special forces operatives linked to stand-off weapons dozens or hundreds of miles away can help national military forces simultaneously understand local context and retain the ability to deliver devastating attacks when necessary (Gray 2005, 251).

It is possible that the information age will continue empowering nonstate actors. The skillful exploitation of the Internet by radical jihadis, ranging from Al Qaeda message boards to Iraqi Sunni recruiting and training Web pages, shows how information technology can already substitute for physical meeting space for nonstate actors (Coll and Glasser 2005).[10] Both Hamas and Hezbollah, for example, have used commercial sources as innocuous as Google Earth to search for targets in Israel and then aim their rockets (Chassey and Johnson 2007). This is especially true when the targeted installations are not official military sites with hidden locations.

As the cost of more sophisticated communications technology declines, the barriers to effective surveillance may increase—the volume of information may overload nearly all intelligence efforts. Hammes and others writing about the impact of information technology on asymmetrical warfare strategies argue that these changes are also decreasing the costs to forming terrorist groups or planning attacks (Hammes 2004, 197–200; Vickers and Martinage 2004, 187–88). Ambassador Hank Crumpton (2006), the former coordinator for counterterrorism at the U.S. State Department, testified about the way that the Internet has already influenced nonstate actors:

Enemy safe havens also include cyberspace. Terrorists often respond to our collective success in closing physical safe havens by fleeing to cyberspace where they seek a new type of safe haven. Harnessing the Internet's potential for speed, security, and global linkage, terrorists increase their ability to conduct some of the activities that once required physical safe haven. They not only use cyberspace to communicate, but also to collect intelligence, disseminate propaganda, recruit operatives, build organizations, fundraise, plan, and even train.

---

[10] Groups can already form and coordinate over the Web and use draft messages at ever-changing free email addresses to communicate in an attempt to evade the electronic surveillance of phones and physical territory (Coll and Glasser 2005).

If cyberattacks become an increasingly effective means of disrupting the activities of military organizations, the Internet could provide a major relative edge to nonstate actors interested in asymmetrical operations. The Internet could function as a virtual leveler that cuts into the ability of nation-states to maintain a monopoly on organized violence in their territories, and helps insurgent and terrorist groups survive. The culmination of these developments is potentially what Audrey Cronin (2006, 77, 86) calls a new age of cybermobilization, where the ability to communicate online could generate nonstate opposition movements run by people from totally different geographic places and societies.

Using the Internet for offensive purposes, through hacking, denial of service attacks, and virus transmissions, instead of as a substitute for physical safe havens is another level of information age activity that would require greater training and capabilities. One need only look at the financial and political consequences of Nigerian email scams to see the potential for harassment/disruption/damage available at low cost to anyone with a computer (Dixon 2005; Olukoya 2002). But the unit cost of those actions are also relatively low; the innocuous computer equipment possessed by nearly all those arrested in the last few years for distributing viruses through cyberspace shows the danger in this regard (Arquilla, Ronfeldt, and Zanini 1999; Glasser and Coll 2005).

There is uncertainty about specific implications, but one can envision insurgent groups organized online and conducting cyberoperations as their primary means of attack. It is also likely that preexisting violent nonstate groups will continue to rely on kinetic violence to achieve their ends. For them, the Internet will be a means of organizing attacks and attempting to defeat electronic surveillance, but their primary focus will not shift into cyberspace (McAllister 2004). Adoption-capacity theory predicts that younger terrorist groups less wedded to traditional brick-and-mortar forms of organizing and carrying out attacks will find it easier to transition to a more complete online presence than more established groups.

At present, Al Qaeda appears to have successfully utilized the information age to escape what otherwise might have been catastrophic defeat in late 2001. Steve Coll and Susan Glasser (2005) write about the shift in Al Qaeda's operational core following the U.S.-led invasion of Afghanistan in fall 2001: "With laptops and DVDs, in secret hideouts and at neighborhood Internet cafes, young code-writing Jihadists have sought to replicate the training, communication, planning and preaching facilities they lost in Afghanistan with countless new locations on the Internet." As chapter 6 describes, Al Qaeda had a high level of organizational capital prior to 2001, meaning in theory it would be capable of making an organizational transition. The level of organizational capital required for adoption of cyberwarfare is likely to be medium or high, given that at a minimum it will mean a large-scale shift in the organizational emphasis of nonstate groups from brick-and-mortar locations to cyberspace, but beyond that it is uncertain.

Finally, it is important to stress yet again that these predictions for state and nonstate actors about the way particular trends will influence the future, especially with regard to military power, are rarely accurate—and are accurate for the right reasons even less frequently. But this exercise shows how adoption-capacity theory is a potentially useful tool not just for academics studying international relations but for those interested in the future of warfare as well.

The Final Word

This book shows the critical importance of understanding how military innovations diffuse and affect international politics. Strategic competition, domestic politics, and international norms are all relevant factors influencing the international system and spread of military power. By focusing on the *ability* of states to successfully adapt in different situations, however, adoption-capacity theory provides a more complete picture of change in international politics, supplying a new answer to the puzzle of how military innovations influence the international security environment. Different adoption requirements cause patterns that reverberate throughout the international system differently. Sometimes, due to its adoption requirements, the way an innovation spreads will reinforce existing power balances, and help major powers win wars faster and with fewer casualties. At other times, the spread of a new military innovation creates a strategic moment in which major power transitions can occur.

Mismatches between the adoption requirements of a new innovation and the capabilities of a great power can help explain the timing and extent of the decline of great powers in some cases. In those conditions, to avoid conflict, great powers may severely alter their grand strategies and voluntarily shift to a much less active status. Alternatively, as sometimes happens, great powers can continue to pursue policy courses that outstrip their actual relative military power, as their attempt at adoption fails. Sometimes states simply lack the assets to adopt the next crucial military innovation. Even if they play their hands in an optimal fashion, they will be unable to maintain their status. It is in this situation that flawed assessments and informational gaps can produce dangerous escalation dynamics, increasing the risk of international conflict.

No one wonders why the German Army rolled through Belgium in 1940. The relative power mismatch meant that the Belgians would have lost even if they had maxed out their capacity. The same can easily be true for other more powerful actors: the characteristics that serve you well in one era may not translate to the next. This is a simpler, clearer, and more accurate explanation for the military rise and decline of states than that provided by international relations theories alone.

SUICIDE TERRORISM GROUP LINKAGES

Table A.1
Group Linkages

Name	Location	Initial link	Nature of link	Citation	Grouped	Grouped with
Abu Hafs Al-Masri Brigade	Europe	Al Qaeda	Direct	Terrorism Knowledge Base 2006	1	AQ-Europe
Al-Aarifeen	Pakistan	Al Qaeda	Indirect	Terrorism Knowledge Base 2006	1	AQ-inspired—Pakistan
Al-Aqsa Martyrs Brigade	Israel	Fatah	Direct	Terrorism Knowledge Base 2006	1	Fatah
Al-Bara Bin Malek Brigades	Iraq	Al Qaeda	Direct	Terrorism Knowledge Base 2006	1	AQ-Iraq
Fatah	Israel	Hamas; Palestinian Islamic Jihad	Indirect	Bloom 2005; Pape 2005; Pedahzur 2005	0	
Al-Islambouli Brigades of Al Qaeda	Russia; Pakistan	Al Qaeda; Riyad us-Saliheyn Martyrs' Brigade	Direct	Terrorism Knowledge Base 2006	0	
Al-Mansoorain	Kashmir	Al Qaeda	Indirect	Terrorism Knowledge Base 2006	1	AQ-inspired—Pakistan
Al Qaeda	World	Hezbollah	Direct	National Commission on Terrorist Attacks upon the United States 2004	0	

Name	Location	Initial link	Nature of link	Citation	Grouped	Grouped with
Al Qaeda in the Arabian Peninsula	Saudi Arabia	Al Qaeda	Direct	Terrorism Knowledge Base 2006	1	AQ_ME
Al Qaeda organization in the Land of the Two Rivers	Middle East	Al Qaeda	Direct	Terrorism Knowledge Base 2006	1	AQ_Iraq
Al-Qanoon	Pakistan	Al Qaeda; Lashkar-e-Jhangvi	Direct	Terrorism Knowledge Base 2006	1	AQ_inspired— Pakistan
Al-Quds Brigades	Israel	Palestinian Islamic Jihad	Direct	Ricolfi 2005	1	Palestinian Islamic Jihad
Amal	Lebanon	Hezbollah	Direct	Kramer 1990	0	
Ansar Al-Islam	Iraq	Al Qaeda	Direct	Terrorism Knowledge Base 2006	1	AQ_Iraq
Ansar Allah	Lebanon	Hezbollah	Direct	Terrorism Knowledge Base 2006	0	
Ansar al-Sunnah Army	Iraq	Al Qaeda	Direct	Terrorism Knowledge Base 2006	1	AQ_Iraq
Armed Islamic Group	Algeria	None			0	
Army for the Liberation of Kurdistan	Lebanon	None			0	
Black Widows	Chechnya/ Russia	Riyad us-Saliheyn Martyrs' Brigade; Al Qaeda	Direct	Pedahzur 2005; Terrorism Knowledge Base 2006	1	Chehnyan

Group	Location	Affiliation	Type	Source		Classification
Revolutionary People's Liberation Party/Front	Turkey	Kurdistan Workers' Party	Indirect	Terrorism Knowledge Base 2006	0	
Hamas	Israel	Hezbollah	Direct	Aboul-Enein 2005, 9; Dolnik and Bhattacharjee 2002, 109; Levitt 2004; Ricolfi 2005, 91–92; Schweitzer 2002, 3	0	
Harakat ul-Mujahidin	Kashmir	Al Qaeda	Direct	Terrorism Knowledge Base 2006	1	AQ inspired—Pakistan
Hezbollah	Lebanon	Iranian government	Direct	Kramer 1990	0	
Hizbul Mujahideen	Kashmir	Lashkar-e-Toiba	Direct	Terrorism Knowledge Base 2006	1	AQ inspired—Pakistan
Iraqi Liberation Army	Lebanon	None			0	
Islami Chhatra Shibir	Bangladesh	Al Qaeda-related groups	Direct	Terrorism Knowledge Base 2006	1	AQ inspired—Bangladesh
Islamic Army in Iraq	Iraq	Al Qaeda-related groups	Direct	Terrorism Knowledge Base 2006	1	AQ Iraq
Islamic Glory Brigades in the Land of the Nile	Egypt	Al Qaeda	Indirect	Terrorism Knowledge Base 2006	1	AQ ME
Islamic Jihad Group	Uzbekistan	Islamic Movement of Uzbekistan; Al Qaeda	Direct	Pedahzur 2005; Terrorism Knowledge Base 2006	1	Uzbekistan
Jamatul Mujahedin Bangladesh	Bangladesh	Al Qaeda-inspired; Islami Chhatra Shibir	Indirect	Terrorism Knowledge Base 2006	1	AQ inspired—Bangladesh

TABLE A.1 (continued)

Name	Location	Initial link	Nature of link	Citation	Grouped	Grouped with
Jemaah Islamiya (JI)	Indonesia	Al Qaeda	Direct	Hoffman and McCormick 2004	0	
Jenin Martyrs' Brigade	Israel	Hamas	Direct	Terrorism Knowledge Base 2006	1	Hamas
Jihad Pegah	Iraq	Al Qaeda	Indirect	Terrorism Knowledge Base 2006	1	AQ_Iraq
Jund Al-Sham (Army of the Levant)	Lebanon; Qatar	Al Qaeda	Direct	Terrorism Knowledge Base 2006	1	AQ_ME
Kurdistan Workers' Party	Turkey			Pedahzur 2005	0	
Lashkar-e-Jhangvi	Pakistan	Al Qaeda; Tailban	Direct	Pedahzur 2005; Terrorism Knowledge Base 2006	1	AQ inspired—Pakistan
Lebanese National Resistance Front	Lebanon	Hezbollah	Indirect	Terrorism Knowledge Base 2006	0	
Liberation Tigers of Tamil Eelam	Sri Lanka	Hezbollah	Direct	Hoffman and McCormick 2004; Pape 2005	0	
Mujahideen Shura Council	Iraq	Al Qaeda	Direct	Terrorism Knowledge Base 2006	1	AQ_Iraq
Palestinian Islamic Jihad	Israel	Hezbollah	Direct	Aboul-Enein 2005, 9; Dolnik and Bhattacharjee 2002, 109; Levitt 2004; Ricolfi 2005, 91–92; Schweitzer 2002, 3	0	

Organization	Location	Affiliation	Relationship	Source		Code
People's Liberation Army of Kurdistan	Turkey	Kurdistan Workers' Party	Direct	Terrorism Knowledge Base 2006	0	
Popular Front for the Liberation of Palestine	Israel	Hamas; Palestinian Islamic Jihad	Indirect	Bloom 2005; Pedahzur 2005	0	
Riyad us-Saliheyn Martyrs' Brigade	Chechnya/ Russia	Al Qaeda	Direct	Pedahzur 2005; Terrorism Knowledge Base 2006	1	AQ_Chechnya
Secret organization of Al Qaeda in Europe	Europe	Al Qaeda	Direct	Terrorism Knowledge Base 2006	1	AQ_Europe
Soldiers of the Prophet's Companions	Iraq	Al Qaeda	Indirect	Terrorism Knowledge Base 2006	1	AQ_Iraq
Syrian inspired Assorted	Lebanon	Hezbollah	Indirect	Kramer 1990	0	
Taliban	Afghanistan	Al Qaeda	Direct	Hoffman and McCormick 2004; Terrorism Knowledge Base 2006	0	
Tanzim	Israel	Fatah	Direct	Terrorism Knowledge Base 2006	1	Fatah
Tawhid and Jihad	Iraq	Al Qaeda	Direct	Terrorism Knowledge Base 2006	1	AQ_Iraq

NUCLEAR DIFFUSION SURVIVAL MODEL

A final check on the results from the nuclear weapons chapter is conducted using an event history, or survival, model. Survival models are especially useful for a task like measuring the acquisition of a weapons system because they are designed to predict failure rates, or the length of time until a given event occurs.[1] Survival models are also helpful for dealing with rare events such as the acquisition of nuclear weapons (less than fifteen states have acquired nuclear weapons over a sixty-one-year period), where the passage of time matters in predicting outcomes (Box-Steffensmeier and Jones 2004; Singh and Way 2004, 871). Creating a survival model involves outlining a "hazard" rate based on the underlying distribution of risk, or the probability that as time passes an event will occur—in this case, the initiation of a nuclear weapons program or acquisition of a nuclear weapon. This makes survival models well suited for measuring something like nuclear proliferation.

Unfortunately, as explained above, the multistage nature of acquiring nuclear weapons is not something that survival models can easily measure, making it necessary to run independent survival models for each stage of the process. Given that a cohesive selection survival model is not available, the survival models are used as a robustness check rather than as the main analytic tool. Building on prior survival analysis on the spread of nuclear weapons by Singh and Way, the appropriate description of the underlying hazard function for the spread of nuclear weapons is a parametric Weibull distribution. One difference in the dependent variables for survival models is that they only register movement when something changes. In other setups, the United States is coded as a nuclear power beginning in 1945, meaning that the United States would register a 1 for a nuclear weapons variable in 1945 and then again in 1946, and so on. In the survival setup, it is only during the year of acquisition that the variable changes. So the United States is scored a 1 in 1945 but not in 1946.

Table A.2 imports the three-tiered Singh-Way coding scheme for levels of nuclear weapons proliferation. The first level is the initiation of a nuclear weapons program, the second level is the active pursuit of a nuclear weapons capability and the third level is the acquisition of nuclear weapons. The results, with one survival model for each stage, replicate the Singh-Way analysis, but insert the financial intensity variables that adoption-capacity theory predicts

---

[1] They were originally developed for predicting patient behavior in the medical field.

TABLE A.2
Survival Analysis of Nuclear Weapons Diffusion Based on Singh-Way Results

	DV: Nuclear program, no controls	DV: Nuclear program, controls	DV: Nuclear pursuit, no controls	DV: Nuclear pursuit, controls	DV: Nuclear acquisition, no controls	DV: Nuclear acquisition, controls
	Coefficient/(robust standard error)					
GDP per capita	0.0001 (0.0002)	0.004 (0.0003)	0.0001 (0.0003)	0.001 (0.0003)***	-0.0001 (0.0003)	0.0002 (0.001)
GDP squared	-1.62E-08 (1.32E-08)	-3.95E-08 (2.20E-08)*	-3.54E-08 (2.14E-08)*	-8.93E-08 (3.21E-08)***	-1.76E-08 (1.80E-08)	-4.91E-08 (3.99E-08)
Industrial capacity	-25.887 (34.556)	-19.593 (25.478)	36.835 (15.938)**	14.436 (19.292)	43.225 (16.999)**	47.935 (18.284)***
Latent nuclear capacity	0.460 (0.124)***	0.400 (0.182)**	0.308 (0.138)**	0.210 (0.197)	0.840 (0.329)**	1.117 (0.619)*
Military spending	1.89E-07 (4.88E-08)***	1.09E-07 (8.04E-08)	1.82E-07 (4.10E-08)***	1.39E-07 (5.50E-08)**	1.93E-07 (6.75E-08)***	3.42E-07 (1.00E-07)***
Organizational age	0.019 (0.013)	0.036 (0.014)**	0.003 (0.021)	0.025 (0.025)	0.049 (0.026)*	0.026 (0.039)
Diffusion	-2.101 (0.536)***	-2.427 (0.797)***	-3.825 (1.894)**	-3.071 (3.684)	-9.473 (7.584)	-20.904 (6.688)***
Enduring rivalry		1.996 (0.642)***		1.659 (1.043)		1.578 (2.197)
Dispute involvement		0.115 (0.072)		-0.33 (0.103)***		0.012 (0.246)
Alliances		-0.895 (0.536)*		-0.927 (0.862)		-2.798 (2.137)

TABLE A.2 (continued)

	DV: Nuclear program, no controls	DV: Nuclear program, controls	DV: Nuclear pursuit, no controls	DV: Nuclear pursuit, controls	DV: Nuclear acquisition, no controls	DV: Nuclear acquisition, controls
Democracy		0.028		0.052		-0.190
		(0.042)		(0.040)		(0.062)***
Democratization		-0.052		-0.1114		0.140
		(0.099)		(0.117)		(0.252)
Percentage of democracies		-0.003		-0.091		-0.013
		(0.072)		(0.128)		(0.128)
Economic openness		-0.007		-0.015		0.042
		(0.014)		(0.017)		(0.024)*
Economic liberalization		-0.040		0.034		-0.025
		(0.022)*		(0.017)**		(0.030)
Constant	0.194	-1.370	-1.939	-5.715	-3.010	9.083
	(1.678)	(3.506)	(3.100)	(6.569)	(10.220)	(8.389)
/ln_p	-0.137	0.158	1.168	1.227	2.083	2.720
	(0.251)	(0.268)***	(0.358)***	(0.561)**	(0.644)***	(0.302)***
p	0.872	1.171	3.216	3.410	8.031	15.182
	(0.219)	(0.313)	(1.119)	(1.912)	(5.172)	(4.585)
1/p	1.146	0.854	0.311	0.293	0.125	0.066
	(0.288)	(0.229)	(0.108)	(0.164)	(0.080)	(0.020)

Notes: Model 1: N: 4042, Wald chi2(7): 56.44, Prob > chi2: 0, Subjects: 148, Wald chi2(7): 56.44, Prob > chi2: 0, Subjects: 148, Failures: 21, Time at risk: 4089, log pseudolikelihood: −59.8954, Cluster adjustment: 148. Model 2: N: 3811, Wald chi2(15): 84.10, Prob > chi2: 0, Subjects: 144, Failures: 20, Time at risk: 3828, Log pseudolikelihood: −43.8572, Cluster adjustment: 144. Model 3: N: 4350, Wald chi2(7): 66.83, Prob > chi2: 0, Subjects: 149, Failures: 15, Time at risk: 4397, Log pseudolikelihood: −37.4771, Cluster adjustment: 149. Model 4: N: 4116, Wald chi2(15) 178.69, Prob > chi2: 0, Subjects: 145, Failures: 14, Time at risk: 4133, Log pseudolikelihood: −21.1119, Cluster adjustment: 145. Model 5: N: 4507, Wald chi2(7): 50.51, Prob > chi2: 0, Subjects: 150, Failures: 9, Time at risk: 4554, Log pseudolikelihood: −13.73606, Cluster adjustment: 150. Model 6: N: 4272, Wald chi2(15) 650.02, Prob > chi2: 0, Subjects: 146, Failures: 9, Time at risk: 4289, Log pseudolikelihood: −4.449215, Cluster adjustment: 146.

* $p < .10$ ** $p < .05$ *** $p < .01$.

will influence the probability of proliferation. The financial intensity variables are significant in a similar pattern to table 4.2, with industrial capacity, in particular, substantively producing a massive increase in the hazard rate in the acquisition category.[2] This check demonstrates the utility of the financial intensity measures across multiple types of models for measuring the spread of nuclear weapons and provides a general robustness check for the findings in chapter 4.

[2] Please contact the author for more detailed results at horom@sas.upenn.edu.

BIBLIOGRAPHY

Abernathy, William J., and Kenneth Wayne. 1974. Limits of the learning curve. *Harvard Business Review* 52 (5): 109–19.

Aboul-Enein, Youssef. 2005. Hamas: A further exploration of jihadist tactics. *Strategic Insights* 4 (9): 12.

Abrahms, Max. 2006. Why terrorism does not work. *International Security* 31 (2): 42–78.

Achen, Christopher H. 2002. Toward a new political methodology: Microfoundations and ART. *Annual Review of Political Science* 5:423–50.

———. 2005. Let's put garbage-can regressions and garbage-can probits where they belong. *Conflict Management and Peace Science* 22 (4): 327–39.

Adams, Christopher. 2007. MoD gives nod for aircraft carriers. *Financial Times*, July 26.

Albright, David. 1994. South Africa and the affordable bomb. *Bulletin of the Atomic Scientists* 50 (4): 37–47.

Anderton, Charles H. 1992. Towards a mathematical theory of the offense/defense balance. *International Studies Quarterly* 36 (1): 75–99.

Arai, Mahmood. 2003. Wages, profits, and capital intensity: Evidence from matched worker-firm data. *Journal of Labor Economics* 21 (3): 593–618.

Arquilla, John. 2003. Patterns of commercial diffusion. In Goldman and Eliason 2003a, 348–70.

Arquilla, John, and David F. Ronfeldt. 2000. *Swarming and the future of conflict*. Santa Monica, CA: RAND.

———, eds. 2001. *Networks and netwars: The future of terror, crime, and militancy*. Santa Monica, CA: RAND.

Arquilla, John, David Ronfeldt, and Michele Zanini. 1999. Networks, netwar, and information-age terrorism. In *Countering the new terrorism*, ed. Ian O. Lesser, Bruce Hoffman, John Arquilla, David Ronfeldt, Michele Zanini, and Brian Michael Jenkins, 39–84. Santa Monica, CA: RAND.

Arreguin-Toft, Ivan. 2005. *How the weak win wars: A theory of asymmetric conflict*. New York: Cambridge University Press.

Asal, Victor, and R. Karl Rethemeyer. 2008. The nature of the beast: Organizational structures and the lethality of terrorist attacks. *Journal of Politics* 70 (2): 437–49.

Atran, Scott. 2003. Genesis of suicide terrorism. *Science* 299 (5612): 1534–39.

Avant, Deborah D. 1994. *Political institutions and military change: Lessons from peripheral wars*. Ithaca, NY: Cornell University Press.

Axelrod, Robert M. 1979. The rational timing of surprise. *World Politics* 31 (2): 228–46.

Ayers, Nick. 2006. Choosing to die: Repression and the adoption of suicide tactics. Unpublished manuscript, Stanford University.

Bacevich, Andrew. 1986. *The pentomic era: The US Army between Korea and Vietnam.* Washington, DC: National Defense University Press.

Bacon, Reginald H. 1929. *The life of Lord Fisher of Kilverstone, admiral of the fleet.* Vol. 1. Garden City, NY: Doubleday, Doran and Company.

Bagley, Worth H., and Gene R. LaRocque. 1976. Superpowers at sea: A debate. *International Security* 1 (1): 56–76.

Bain, Joe S. 1956. *Barriers to new competition: Their character and consequences in manufacturing industries.* Cambridge, MA: Harvard University Press.

Baldwin, David A. 1979. Power analysis and world politics: New trends versus old tendencies. *World Politics* 31 (2): 161–94.

Barnhart, Michael A. 1999. Book review: Evans and Peattie, Kaigun: Strategy, tactics, and technology in the Imperial Japanese Navy, 1887–1941. *Journal of Japanese Studies* 25 (1): 211–14.

Beck, Nathaniel, Jonathan N. Katz, and Richard Tucker. 1998. Taking time seriously: Time-series-cross-section analysis with a binary dependent variable. *American Journal of Political Science* 42 (4): 1260–88.

Bell, J. Bowyer. 1990. *IRA tactics and targets.* Dublin: Poolbeg.

———. 1998. *The secret army: The IRA.* Rev. 3rd ed. Dublin: Poolbeg.

Benjamin, Daniel, and Steven Simon. 2002. *The age of sacred terror.* New York: Random House.

Benjamin, Mark. 2008. Killing "Bubba" from the skies. Salon.com. February 15. Available at http://www.salon.com/news/feature/2008/02/15/air_war/index.html.

Bennett, D. Scott, and Allan C. Stam. 1996. The duration of interstate wars, 1816–1985. *American Political Science Review* 90 (2): 239–57.

———. 2000. EUGene: A conceptual manual. *International Interactions* 26 (2): 179–204.

———. 2004. *The behavioral origins of war.* Ann Arbor: University of Michigan Press.

Benthem van den Bergh, Godfried. 1992. *The nuclear revolution and the end of the cold war: Forced restraint.* Houndmills, UK: Macmillan Press.

Bergen, Peter L. 2002. *Holy war, inc.: Inside the secret world of Osama bin Laden.* 1st Touchstone ed. New York: Simon and Schuster.

Berman, Eli, John Bound, and Stephen Machin. 1998. Implications of skill-biased technological change: International evidence. *Quarterly Journal of Economics* 113 (4): 1245–79.

Berman, Eli, and David D. Laitin. 2005. Hard targets: Theory and evidence on suicide attacks. *National Bureau of Economic Research Working Paper* 11740 (November). Available at http://ssrn.com/abstract=842475.

Berry, Frances S., and William D. Berry. 1990. State lottery adoptions as policy innovations: An event history analysis. *American Political Science Review* 84 (2): 395–415.

———. 1992. Tax innovation in the states: Capitalizing on political opportunity. *American Journal of Political Science* 36 (3): 715–42.

Betts, Richard K. 1987. *Nuclear blackmail and nuclear balance.* Washington, DC: Brookings Institution.

Biddle, Stephen D. 2004. *Military power: Explaining victory and defeat in modern battle.* Princeton, NJ: Princeton University Press.

———. 2007a. Explaining military outcomes. In *Creating military power: The sources of military effectiveness,* ed. Risa A. Brooks and Elizabeth A. Stanley, 207–27. Stanford, CA: Stanford University Press.

———. 2007b. Strategy in war. *PS: Political Science and Politics* 40 (3): 461–66.

Biddle, Stephen D., and Jeffrey A Friedman. 2008. *The 2006 Lebanon campaign and the future of warfare: Implications for army and defense policy.* Carlisle Barracks, PA: Strategic Studies Institute, U.S. Army War College.

Bin Laden, Osama. 1996. Declaration of war against the Americans occupying the land of the two holy places. *Al Quds Al Arabi,* August. Available at http://www.pbs.org/newshour/terrorism/international/fatwa_1996.html.

Bishop, Chris, and Chris Chant. 2004. *Aircraft carriers: The world's greatest naval vessels and their aircraft.* Saint Paul, MN: Summertime Publishing Ltd.

Blais, Andre, and Louis Massicotte. 1997. Electoral formulas: A macroscopic perspective. *European Journal of Political Research* 32 (1): 107–29.

Bloom, Mia. 2005. *Dying to kill: The allure of suicide terror.* New York: Columbia University Press.

Bolkcom, Christopher. 2005. F/A-22 Raptor. Congressional Research Service. March 3. Available at http://digital.library.unt.edu/govdocs/crs/permalink/meta-crs-6344:1.

Bond, Stephen R., and Jason G. Cummins. 2000. The stock market and investment in the new economy: Some tangible facts and intangible fictions. *Brookings Papers on Economic Activity* 2000 (1): 61–124.

Boot, Max. 2006. *War made new: Technology, warfare, and the course of history, 1500 to today.* New York: Gotham Books.

Borensztein, Eduardo, Jose De Gregorio, and Jong-Wa Lee. 1998. How does foreign direct investment affect economic growth? *Journal of International Economics* 45 (1): 115–35.

Botti, Timothy J. 1996. *Ace in the hole: Why the United States did not use nuclear weapons in the cold war, 1945 to 1965.* Westport, CT: Greenwood Press.

Box-Steffensmeier, Janet M., and Bradford S. Jones. 2004. *Event history modeling: A guide for social scientists.* New York: Cambridge University Press.

Bradley, Omar N. 1967. *The collected writings of General Omar N. Bradley.* Vol. 1. Washington, DC.

Brassey's. 2004. *Imperial hubris: Why the West is losing the war on terror.* Washington, DC: Brassey's.

Braumoeller, Bear F. 2004. Hypothesis testing and multiplicative interaction terms. *International Organization* 58 (4): 807–20.

Bremer, Stuart A. 1982. The contagiousness of coercion: The spread of serious international disputes, 1900–1976. *International Interactions* 9 (1): 29–55.

Breyer, Siegfried. 1970. *Guide to the Soviet Navy*. Trans. L.C.M. Henley. 1st ed. Annapolis, MD: U.S. Naval Institute.

———. 1973. *Battleships and battle cruisers, 1905–1970*. Trans. Alfred Kurti. Garden City, NY: Doubleday.

Brodie, Bernard. 1941. *Sea power in the machine age*. Princeton, NJ: Princeton University Press.

———. 1946. *The absolute weapon: Atomic power and world order*. New York: Harcourt Brace.

Brodie, Bernard, and Fawn McKay Brodie. 1973. *From crossbow to H-bomb*. Bloomington: Indiana University Press.

Brooks, John. 2005. *Dreadnought gunnery and the battle of Jutland: The question of fire control*. New York: Routledge.

Brooks, Risa A. 2007. Conclusion. In *Creating military power: The sources of military effectiveness*, ed. Risa A. Brooks and Elizabeth A. Stanley, 228–37. Stanford, CA: Stanford University Press.

Brooks, Stephen G., and William C. Wohlforth. 2008. *World out of balance: International relations and the challenge of American primacy*. Princeton, NJ: Princeton University Press.

Brynjolfsson, Erik, and Loren M. Hitt. 2000. Beyond computation: Information technology, organizational transformation, and business performance. *Journal of Economic Perspectives* 14 (4): 23–48.

Brynjolfsson, Erik, Lorin M. Hitt, and Shinkyu Yang. 2002. Intangible assets: Computers and organizational capital. *Brookings Papers on Economic Activity* 2002 (1): 137–81.

Bueno de Mesquita, Bruce, James Morrow, and Ethan Zorick. 1997. Capabilities, perception, and escalation. *American Political Science Review* 91 (1): 15–27.

Bueno de Mesquita, Bruce, and William H. Riker. 1982. An assessment of the merits of selective nuclear proliferation. *Journal of Conflict Resolution* 26 (2): 283–306.

Bunn, George. 1994. The NPT and options for its extension in 1995. *Nonproliferation Review* (Winter): 52–60.

Bureau of Aeronautics. 1943. Model-F4F-4. In *Historical Aircraft*. Washington, DC: Naval Historical Center.

Busch, Per-Olof, and Helge Jörgens. 2005. The international sources of policy convergence: Explaining the spread of environmental policy innovations. *Journal of European Public Policy* 12 (5): 860–84.

Bush, George W. 2006. *The national security strategy of the United States of America*. Washington, DC: White House. Available at http://www.whitehouse.gov/nsc/nss/2006/.

Byman, Daniel L. 2003. Al-Qaeda as an adversary: Do we understand our enemy? *World Politics* 56 (1): 139–63.

Canon, Bradley C., and Lawrence Baum. 1981. Patterns of adoption of tort law innovations: An application of diffusion theory to judicial doctrines. *American Political Science Review* 75 (4): 975–87.

Carton, Donna. 2005. Terrorist bomb blitz: It was unlikely we didn't have any suicide bombers; RIRA boss Michael McKevitt. *Sunday Mirror*, July 17, 4–5.

Cederman, Lars-Erik, and Kristian Skrede Gleditsch. 2004. Conquest and regime change: An evolutionary model of the spread of democracy and peace. *International Studies Quarterly* 48 (3): 603–29.

Chassey, Clancy, and Bobbie Johnson. 2007. Google Earth used to target Israel. *Guardian*, October 25. Available at http://www.guardian.co.uk/technology/2007/oct/25/google.israel.

Chayes, Abram, and Antonia H. Chayes. 1995. *The new sovereignty: Compliance with international regulatory agreements.* Cambridge, MA: Harvard University Press.

Chernyavskii, Sergei. 2005. The era of Gorshkov: Triumph and contradictions. *Journal of Strategic Studies* 28 (2): 281–308.

Chesneau, Roger. 1980. *Conway's all the world's fighting ships, 1922–1946.* London: Conway Maritime Press.

———. 1984. *Aircraft carriers of the world, 1914 to the present: An illustrated encyclopedia.* London: Arms and Armour.

Christensen, Clayton M. 1997. *The innovator's dilemma: When new technologies cause great firms to fail.* Boston: Harvard Business School Press.

Cipolla, Carlo M. 1965. *Guns and sails in the early phase of European expansion, 1400–1700.* London: Collins.

Clark, Kim B. 1985. The interaction of design hierarchies and market concepts in technological evolution. *Research Policy* 14:235–51.

Coates, Dennis, and Jac C. Heckelman. 2003. Interest groups and investment: A further test of the Olson hypothesis. *Public Choice* 117 (3): 333–40.

Cohen, Eliot A. 1996. A revolution in warfare. *Foreign Affairs* 75 (2): 37–54.

Cohen, Avner. 1998. *Israel and the bomb.* New York: Columbia University Press.

Coll, Steve, and Susan B. Glasser. 2005. Terrorist turn to the Web as base of operations. *Washington Post*, August 7, A1.

Collins, Eamon, and Mick McGovern. 1997. *Killing rage.* London: Granta Books.

Coogan, Tim Pat. 1993. *The IRA: A history.* 1st U.S. ed. Niwot, CO: Roberts Rinehart Publishers.

Copeland, Dale C. 2000. *The origins of major war.* Ithaca, NY: Cornell University Press.

Correlates of War Project. 2007. *State system membership list, v2004.1.* February 19. Available at http://correlatesofwar.org.

Correlates of War 2 Project. 2006. *National material capabilities data documentation, updated for version 3.2.* January 20. Available at http://cow2.la.psu.edu/.

Cote, Owen R., Jr. 2003. *The third battle: Innovation in the navy's silent cold war struggle with Soviet submarines, Newport paper #16.* Newport, RI: U.S. Naval War College Press.

Coutau-Begarie, Herve. 1996. French naval strategy: A naval power in a continental environment. In *Naval power in the twentieth century*, ed. N.A.M. Rodger, 59–65. London: Macmillan Press, Ltd.

Crane, Keith, Roger Cliff, Evan S. Medeiros, James C. Mulvenon, and William H. Overholt. 2005. *Modernizing China's military: Opportunities and constraints.* Santa Monica, CA: RAND.

Crease (Fisher) Papers. The papers of Thomas E. Crease (Fisher papers). Portsmouth, UK: Royal Naval Museum Library/Admiralty Library.

Cronin, Audrey K. 2006. Cyber-mobilization: The new levée en masse. *Parameters*: 77–87.

Cronin, Blaise, and Holly Crawford. 1999. Raising the intelligence stakes: Corporate information warfare and strategic surprise. *Competitive Intelligence Review* 10 (3): 58–66.

Crowl, Philip A. 1986. Alfred Thayer Mahan: The naval historian. In *Makers of modern strategy: From Machiavelli to the nuclear age*, ed. Peter Paret, 444–77. Princeton, NJ: Princeton University Press.

Crumpton, Henry A. 2006. The changing face of terror: A post 9/11 assessment. Testimony before the Senate Foreign Relations Committee, Washington, DC. Available at http://www.state.gov/s/ct/rls/rm/2006/68608.htm.

Cummins, Jason G. 2004. A new approach to the valuation of intangible capital. FEDS working paper no. 2004-17 (April). Available at http://ssrn.com/abstract=559461.

Dahl, Erik J. 2005. Net-centric before its time: The Jeune École and its lessons for today. *Naval War College Review* 58 (4): 109–35.

Danilovic, Vesna. 2002. *When the stakes are high: Deterrence and conflict among major powers*. Ann Arbor: University of Michigan Press.

Darby, Michael R., and Lynne G. Zucker. 2003. Growing by leaps and inches: Creative destruction, real cost reduction, and inching up. *Economic Inquiry* 41 (1): 1–19.

David, Paul A. 1986. Technology diffusion, public policy, and industrial competitiveness. In *The positive sum strategy: Harnessing technology for economic growth*, ed. Ralph Landau and Nathan Rosenberg, 373–91. Washington, DC: National Academy Press.

Diamond, Andrew F. 2006. Dying with eyes open or closed: The debate over a Chinese aircraft carrier. *Korean Journal of Defense Analysis* 28 (1): 35–38.

Diehl, Paul F. 1985. Contiguity and military escalation in major power rivalries, 1816–1980. *Journal of Politics* 47 (4): 1203–11.

Diehl, Paul F., and Gary Goertz. 2000. *War and peace in international rivalry*. Ann Arbor: University of Michigan Press.

DiMaggio, Paul J., and Walter W. Powell. 1983. The iron cage revisited: Institutional isomorphism and collective rationality in organizational fields. *American Sociological Review* 48 (2): 147–60.

Dixon, Robyn. 2005. The lure of easy money. *Guardian*, November 10, T1.

Dolnik, Adam, and Anjali Bhattacharjee. 2002. Hamas: Suicide bombings, rockets, or WMD? *Terrorism and Political Violence* 14 (3): 109–28.

Dombrowski, Peter J., and Eugene Gholz. 2006. *Buying military transformation: Technological innovation and the defense industry*. New York: Columbia University Press.

Dorgan, Stephen J., and John J. Dowdy. 2004. When IT lifts productivity. *McKinsey Quarterly* (4): 13–15.

Downs, George W., David M. Rocke, and Peter N. Barsoom. 1996. Is the good news about compliance good news about cooperation? *International Organization* 50 (3): 379–406.

Druks, Herbert. 1967. *Harry S. Truman and the Russians, 1945–1953*. New York: R. Speller.

Echevarria, Antulio. 2005. Deconstructing the theory of fourth-generation war. *Contemporary Security Policy* 26 (2): 233–41.

Edgerton, David. 1996. *Science, technology, and the British industrial "decline," 1870–1970*. New York: Cambridge University Press.

Ehrhard, Thomas Paul. 2001. Unmanned aerial vehicles in the United States armed services: A comparative study of weapon system innovation. PhD diss., Johns Hopkins University.

Elman, Colin. 1999. The logic of emulation: The diffusion of military practices in the international system. PhD diss., Columbia University.

Elster, Jon. 2005. Motivations and beliefs in suicide missions. In Gambetta 2005b, 233–58.

English, Richard. 2003. *Armed struggle: The history of the IRA*. New York: Oxford University Press.

Erickson, Andrew S., and Andrew R. Wilson. 2006. China's aircraft carrier dilemma. *Naval War College Review* 59 (4): 12–45.

Evangelista, Matthew. 1988. *Innovation and the arms race: How the United States and the Soviet Union develop new military technologies*. Ithaca, NY: Cornell University Press.

Evans, David C., and Mark R. Peattie. 1997. *Kaigun: Strategy, tactics, and technology in the Imperial Japanese Navy, 1887–1941*. Annapolis, MD: Naval Institute Press.

Eyre, Dana P., and Mark C. Suchman. 1996. Status, norms, and the proliferation of conventional weapons: An institutional theory approach. In *The culture of national security: Norms and identity in world politics*, ed. Peter J. Katzenstein, 79–113. New York: Columbia University Press.

Falk, Edwin A. 1936. *Togo and the rise of Japanese sea power*. New York: Longmans Green and Company.

Farrell, Theo. 1998. Culture and military power. *Review of International Studies* 24 (3): 407–16.

———. 2005. World culture and military power. *Security Studies* 14 (3): 448–88.

Fearon, James D. 1994a. Domestic political audiences and the escalation of international disputes. *American Political Science Review* 88 (3): 577–92.

———. 1994b. Signaling versus the balance of power and interests: An empirical test of a crisis bargaining model. *Journal of Conflict Resolution* 38 (2): 236–69.

———. 1995a. The offense-defense balance and war since 1648. Paper prepared for the annual meeting of the International Studies Association, Chicago.

———. 1995b. Rationalist explanations for war. *International Organization* 49 (3): 379–414.

Fearon, James D., and David D. Laitin. 2003. Ethnicity, insurgency, and civil war. *American Political Science Review* 97 (1): 75–90.

Fieldhouse, Richard W., and Shunji Taoka. 1989. *Superpowers at sea: An assessment of the naval arms race*. New York: Oxford University Press.

Finnemore, Martha. 1996. Norms, culture, and world politics: Insights from sociology's institutionalism. *International Organization* 50 (2): 325–47.

Finnemore, Martha, and Kathryn Sikkink. 1998. Norms and international relations theory. *International Organization* 52 (4): 887–917.

Fischerkeller, Michael P. 1998. David versus Goliath: Cultural judgments in asymmetric wars. *Security Studies* 7 (4): 1–43.

Fisher, John A. 1920. *Memories and records.* Vol. 1. New York: George H. Doran Company.

———. 1952. *Fear God and dread nought: The correspondence of Admiral of the Fleet Lord Fisher of Kilverstone.* Ed. Arthur J. Marder. Vol. 1. Cambridge, MA: Harvard University Press.

———. 1956. *Fear God and dread nought: The correspondence of Admiral of the Fleet Lord Fisher of Kilverstone.* Ed. Arthur J. Marder. Vol. 2. London: Cape.

———. 1960. *The papers of Admiral Sir John Fisher.* Ed. Peter K. Kemp. Vol. 1. London: Navy Records Society.

———. 1964. *The papers of Admiral Sir John Fisher.* Ed. Peter K. Kemp. Vol. 2. London: Navy Records Society.

Fisher, Richard, Jr. 2006. 2005: A turning point for China's aircraft carrier ambitions. In *International assessment and strategy center.* Available at http://www.strategycenter.net/research/pubID.87/pub_detail.asp.

Freedman, Lawrence. 2005. War evolves into the fourth generation: A comment on Thomas X. Hammes. *Contemporary Security Policy* 26 (2): 254–63.

Friedberg, Aaron L. 1988. *The weary titan: Britain and the experience of relative decline, 1895–1905.* Princeton, NJ: Princeton University Press.

Friedman, Norman. 1986. *The postwar naval revolution.* Annapolis, MD: Naval Institute Press.

Fuhrmann, Matthew. 2009. Taking a walk on the supply side: The determinants of civilian nuclear cooperation. *Journal of Conflict Resolution* 53 (2): 181–208.

Fuquea, David C. 1997. Task Force One: The wasted assets of the United States Pacific battleship fleet, 1942. *Journal of Military History* 61 (4): 707–34.

Fyfe, Herbert C. 1902. *Submarine warfare, past, present, and future.* London: G. Richards.

Gaddis, John Lewis. 1997. *We now know: Rethinking cold war history.* New York: Clarendon Press.

Gambetta, Diego. 2005a. Can we make sense of suicide missions? In Gambetta 2005b, 259–99.

———, ed. 2005b. *Making sense of suicide missions.* New York: Oxford University Press.

Gardiner, Robert, ed. 1993. *Navies in the nuclear age: Warships since 1945.* London: Conway Maritime.

Gardiner, Robert, Randal Gray, and Przemyslaw Budzbon. 1985. *Conway's all the world's fighting ships, 1906–1921.* London: Conway Maritime Press.

Gartner, Scott S. 1997. *Strategic assessment in war.* New Haven, CT: Yale University Press.

Gartzke, Erik. 2001. Democracy and the preparation for war: Does regime type affect states' anticipation of casualties? *International Studies Quarterly* 45 (3): 467–84.

Geddes, Barbara. 1999. Authoritarian breakdown: Empirical test of a game theoretic argument. Paper presented at the American Political Science Association annual meeting, Atlanta.

General Accounting Office. 1996. Information security: Computer attacks at Department of Defense pose increasing risks: Report to congressional requesters. General Accounting Office. Available at http://www.gao.gov/archive/1996/ai96084.pdf.

Gentile, Gian P. 2008. A (slightly) better war: A narrative and its defects. *World Affairs Journal* (Summer). Available at http://www.worldaffairsjournal.org/2008%20-%20Summer/full-Gentile.html.

George, Alexander L. 1993. *Bridging the gap: Theory and practice in foreign policy.* Washington, DC: United States Institute of Peace Press.

Gerschenkron, Alexander. 1962. *Economic backwardness in historical perspective: Book of essays.* Cambridge, MA: Harvard University Press.

Ghosn, Faten, and D. Scott Bennett. 2003. *Codebook for the dyadic militarized interstate incident data, version 3.0.* October 10. Available at http://www.correlatesof war.org.

Gibbon, Edward. 1788/1974. *The history of the decline and fall of the Roman Empire.* Vol. 3. Repr., New York: Modern Library.

Gibler, Douglas, and Meredith R. Sarkees. 2004. Measuring alliances: The Correlates of War formal interstate alliance data set, 1816–2000. *Journal of Peace Research* 41 (2): 211–22.

Gilpin, Robert. 1981. *War and change in world politics.* New York: Cambridge University Press.

Glasser, Susan B., and Steve Coll. 2005. The Web as weapon; Zarqawi intertwines acts on ground in Iraq with propaganda campaign on the Internet. *Washington Post,* August 9.

Gleditsch, Nils Petter, and Havard Hegre. 1997. Peace and democracy: Three levels of analysis. *Journal of Conflict Resolution* 41 (2): 283–310.

Goemans, Hein E. 2000. *War and punishment: The causes of war termination and the First World War.* Princeton, NJ: Princeton University Press.

Golder, Peter N., and Gerard J. Tellis. 1997. Will it ever fly? Modeling the takeoff of really new consumer durables. *Marketing Science* 16 (3): 256–70.

Goldman, Emily O. 2003. Receptivity to revolution: Carrier air power in peace and war. In Goldman and Eliason 2003a, 267–303.

———. 2006. Cultural foundations of military diffusion. *Review of International Studies* 32 (1): 69–91.

———. 2007. International competition and military effectiveness: Naval air power, 1919–1945. In *Creating military power: The sources of military effectiveness,* ed. Risa A. Brooks and Elizabeth A. Stanley, 158–85. Stanford, CA: Stanford University Press.

Goldman, Emily O., and Richard Andres. 1999. Systemic effects of military innovation and diffusion. *Security Studies* 8 (4): 79–125.

Goldman, Emily O., and Leslie C. Eliason, eds. 2003a. *The diffusion of military technology and ideas*. Stanford, CA: Stanford University Press.

———. 2003b. Introduction: Theoretical and comparative perspectives on innovation and diffusion. In Goldman and Eliason 2003a, 1–30.

Goldman, Emily O., and Andrew L. Ross. 2003. Conclusion: The diffusion of military technology and ideas—Theory and practice. In Goldman and Eliason 2003a, 371–403.

Gordon, Harvey. 2002. The "suicide" bomber: Is it a psychiatric phenomenon? *Psychiatric Bulletin* 26:285–87.

Gourevitch, Peter A. 1978. The second image reversed: The international sources of domestic politics. *International Organization* 32 (4): 881–912.

Gray, Colin S. 1992. *House of cards: Why arms control must fail, Cornell studies in security affairs*. Ithaca, NY: Cornell University Press.

———. 2002. *Strategy for chaos: Revolutions in military affairs and the evidence of history*. London: Frank Cass.

———. 2005. *Another bloody century: Future warfare*. London: Phoenix Press.

———. 2006. *Strategy and history: Essays on theory and practice*. New York: Routledge.

Gray, Virginia. 1973. Innovation in the states: A diffusion study. *American Political Science Review* 67 (4): 1174–85.

Green, Michael J., and Katsuhisa Furukawa. 2000. New ambitions, old obstacles: Japan and its search for an arms control strategy. *Arms Control Today* 30 (6): 19–20.

Grissom, Adam. 2006. The future of military innovation studies. *Journal of Strategic Studies* 29 (5): 905–34.

Grove, Eric J. 1987. *From Vanguard to Trident: British naval policy since World War II*. Annapolis, MD: Naval Institute Press.

Grove, Eric, and Geoffrey Till. 1989. Anglo-American maritime strategy in the era of massive retaliation, 1945–1960. In *Maritime strategy and the balance of power: Britain and America in the twentieth century*, ed. John B. Hattendorf and Robert S. Jordan, 271–303. London: Macmillan Press Ltd.

Groves, Leslie R. 1962. *Now it can be told: The story of the Manhattan Project*. 1st ed. New York: Harper.

Gunaratna, Rohan. 2002. *Inside Al Qaeda: Global network of terror*. New York: Columbia University Press.

Haas, Peter M. 1992. Introduction: Epistemic communities and international policy coordination. *International Organization* 46 (1): 1–35.

Hall, Peter A., and Rosemary C. R. Taylor. 1996. Political science and the three new institutionalisms. *Political Studies* 44 (5): 936–57.

Halperin, Morton H. 1987. *Nuclear fallacy: Dispelling the myth of nuclear strategy*. Cambridge, MA: Ballinger Publishing Co.

Halpern, Paul G., ed. 1972. *The Keyes papers: Selection from the private and official correspondence of admiral of the fleet baron Keyes of Zeebrugge*. Vol. 1:1914–18, Publications of the Navy Records Society. London: William Clowes and Sons Limited.

———. 1994. *A naval history of World War I*. Annapolis, MD: Naval Institute Press.

Hammarstrom, Mats. 1994. The diffusion of military conflict: Central and South-East Europe in 1919–20 and 1991–92. *Journal of Peace Research* 31 (3): 263–80.

Hammes, Thomas X. 2004. *The sling and the stone: On war in the 21st century*. Saint Paul, MN: Zenith Press.

Hanks, Robert. 1981. *The Pacific Far East: Endangered American strategic position*. Cambridge, MA: Institute for Foreign Policy Analysis.

Hassner, Ron. 2003. To halve and to hold. *Security Studies* 12 (4): 1–33.

Hattendorf, John B., and Robert S. Jordan. 1989. Conclusion: Maritime strategy and national policy: Historical accident or purposeful planning? In *Maritime strategy and the balance of power: Britain and America in the twentieth century*, ed. John B. Hattendorf and Robert S. Jordan, 348–55. London: Macmillan Press Ltd.

Hattendorf, John B., Roger J. B. Knight, Alan W. H. Pearsall, Nicholas A. M. Rodger, and Geoffrey Till, eds. 1993. *British naval documents, 1204–1960*. Vol. 131, Publications of the Navy Records Society. Brookfield, VT: Ashgate for the Navy Records Society.

Haulman, Daniel L. 2003. U.S. unmanned aerial vehicles in combat, 1991–2003. In *Air Force Historical Research Agency: Short studies on recent operations*. Available at http://www.au.af.mil/au/afhra/short_studies/shortstudies.asp.

Henderson, Rebecca M. 1993. Underinvestment and incompetence as responses to radical innovation: Evidence from the photolithographic alignment equipment industry. *RAND Journal of Economics* 24 (2): 248–70.

Henderson, Rebecca M., and Kim B. Clark. 1990. Architectural innovation: The reconfiguration of existing product technologies and the failure of established firms. *Administrative Science Quarterly* 35 (1): 9–30.

Hendrix, Henry J., II. 2005. An unlikely location. *Naval History*. Available at http://www.military.com/NewContent/0,13190,NH_0705_Location-P1,00.html.

Henisz, Withold J., Bennet A. Zelner, and Mauro F. Guillén. 2005. The worldwide diffusion of market-oriented infrastructure reform, 1977–1999. *American Sociological Review* 70 (6): 871–97.

Herrera, Geoffrey L. 2006. *Technology and international transformation: The railroad, the atom bomb, and the politics of technological change*. Albany: State University of New York Press.

Herrera, Geoffrey L., and Thomas G. Mahnken. 2003. Military diffusion in nineteenth-century Europe: The Napoleonic and Prussian military systems. In Goldman and Eliason 2003a, 205–42.

Herrick, Robert W. 1968. *Soviet naval strategy*. Annapolis, MD: U.S. Naval Institute.

Herwig, Holger H. 1987. *"Luxury" fleet: The Imperial German Navy, 1888–1918*. Rev. ed. London: Ashfield Press.

———. 1997. *The First World War: Germany and Austria-Hungary, 1914–1918*. New York: St. Martin's Press.

———. 2001. The battlefleet revolution, 1885–1914. In *The dynamics of military revolution: 1300–2050*, ed. MacGregor Knox and Williamson Murray, 114–31. New York: Cambridge University Press.

Heston, Alan, Robert Summers, and Bettina Aten. 2002. *Penn world table version 6.1.* Philadelphia: University of Pennsylvania, Center for International Comparisons.

Heyman, Edward, and Edward Mickolus. 1980. Observations on "why violence spreads." *International Studies Quarterly* 24 (2): 299–305.

Hill, J. Richard 1996. British naval planning post-1945. In *Naval power in the twentieth century*, ed. Nicholas A. M. Rodger, 215–26. London: Macmillan Press Ltd.

Hobson, Rolf. 2002. *Imperialism at sea: Naval strategic thought, the ideology of sea power, and the Tirpitz plan, 1875–1914.* Boston: Brill Academic Publishers.

Hoffman, Bruce. 1998. *Inside terrorism.* New York: Columbia University Press.

———. 2002. Lessons of 9/11. Submitted for the committee record to the United States joint September 11, 2001 inquiry staff of the House and Senate Select Committees on Intelligence. October 8. Available at http://www.rand.org/pubs/testimonies/2005/CT201.pdf.

———. 2003. The logic of suicide terrorism. *Atlantic Monthly* 291 (5): 40–47.

Hoffman, Bruce, and Gordon H. McCormick. 2004. Terrorism, signaling, and suicide attack. *Studies in Conflict and Terrorism* 27 (4): 243–81.

Hoffman, Frank G. 2007. *Conflict in the 21st century: The rise of hybrid wars.* Arlington, VA: Potomac Institute for Policy Studies.

———. 2008. *From preponderance to partnership: American maritime power in the 21st century.* Washington, DC: Center for a New American Security.

Holloway, David. 1994. *Stalin and the bomb: The Soviet Union and atomic energy, 1939–1956.* New Haven, CT: Yale University Press.

Holmes, Stephen. 2005. Al-Qaeda, September 11, 2001. In Gambetta 2005b, 131–72.

Hone, Thomas C. 1998. Letters to the editor. *Journal of Military History* 62 (2): 471–73.

Hone, Thomas C., Norman Friedman, and Mark D. Mandeles. 1999. *American and British aircraft carrier development, 1919–1941.* Annapolis, MD: Naval Institute Press.

Hopgood, Stephen. 2005. Tamil Tigers, 1987–2002. In Gambetta 2005b, 43–76.

Horowitz, Michael C. 2008. A tale of two innovations: The 1904 fleet redeployment and the British experience with battlefleet warfare and flotilla defense. Unpublished paper, University of Pennsylvania.

———. 2009. The spread of nuclear weapons and international conflict: Does experience matter? *Journal of Conflict Resolution* 53 (2): 234–57.

Horowitz, Michael C., Rose McDermott, and Allan C. Stam. 2005. Leader age, regime type, and violence international conflict. *Journal of Conflict Resolution* 49 (5): 661–85.

Horowitz, Michael C., Erin Simpson, and Allan C. Stam. 2009. *Domestic institutions and wartime casualties.* Unpublished paper, University of Pennsylvania.

Horton, Roy E. 2000. *Out of (South) Africa: Pretoria's nuclear weapons experience, ACDIS occasional paper.* Champaign: ACDIS, University of Illinois.

Hudson, George E. 1976a. Soviet naval doctrine and Soviet politics, 1953–1975. *World Politics* 29 (1): 90–113.

———. 1976b. Soviet naval doctrine under Lenin and Stalin. *Soviet Studies* 28 (1): 42–65.

Hudson, Valerie M., and Andrea Den Boer. 2002. A surplus of men, a deficit of peace: Security and sex ratios in Asia's largest states. *International Security* 26 (4): 5–38.

Huntington, Samuel P. 1957. *The soldier and the state: The theory and politics of civil-military relations*. Cambridge, MA: Belknap Press of Harvard University Press.

Huth, Paul K. 1988. *Extended deterrence and the prevention of war*. New Haven, CT: Yale University Press.

Huth, Paul K., and Bruce Russett. 1984. What makes deterrence work? Cases from 1900 to 1980. *World Politics: A Quarterly Journal of International Relations* 36 (4): 496–526.

Hymans, Jacques. 2006. *The psychology of nuclear proliferation: Identity, emotions, and foreign policy*. New York: Cambridge University Press.

Iannaccone, Laurence R. 2006. The market for martyrs. *Interdisciplinary Journal of Research on Religion* 2 (4). Available at http://www.religjournal.com/articles/article_view.php?id=16.

Ienaga, Saburo. 1978. *The Pacific War, 1931–1945: A critical perspective on Japan's role in World War II*. New York: Pantheon Books.

Israeli Defense Forces spokesperson. 2002. Documents: The Palestinian Authority employs Fatah activists involved in terrorism and suicide attacks. Israeli Defense Forces. April 23. Available at http://www.imra.org.il/story.php3?id=11558.

Jackson, Brian A. 2005. Provisional Irish Republican Army. In *Aptitude for destruction: Case studies of organizational learning in five terrorist groups*, ed. Brian A. Jackson John C. Baker, Peter Chalk, Kim Cragin, John V. Parachini, and Horacio R. Trujillo, 93–140. Santa Monica, CA: RAND.

Jackson, Brian A., Peter Chalk, Kim Cragin, Bruce Newsome, John V. Parachini, William Rosenau, Erin M. Simpson, Melanie W. Sisson, and Donald Temple. 2007. *Breaching the fortress wall: Understanding terrorist efforts to overcome defensive technologies*. Santa Monica, CA: RAND Corporation.

Jackson Wade, Sara, and Dan Reiter. 2007. Does democracy matter? *Journal of Conflict Resolution* 51 (2): 329–48.

Jaggers, Keith, and Ted Robert Gurr. 1995. Tracking democracy's third wave with the Polity III data. *Journal of Peace Research* 32 (4): 469–82.

Jane's Information Group, ed. 1935. *Jane's fighting ships*. London: Sampson Low Marston and Co.

———, ed. 1950. *Jane's fighting ships*. London: Sampson Low Marston and Co.

———, ed. 1955. *Jane's fighting ships*. London: Sampson Low Marston and Co.

———, ed. 1977. *Jane's fighting ships*. London: Sampson Low Marston and Co.

———, ed. 2006. *Jane's fighting ships*. London: Sampson Low Marston and Co.

Jervis, Robert. 1976. *Perception and misperception in international politics*. Princeton, NJ: Princeton University Press.

———. 1986. The nuclear revolution and the common defense. *Political Science Quarterly* 101 (5): 689–703.

———. 1988. The political effects of nuclear weapons: A comment. *International Security* 13 (2): 80–90.

————. 1989. *The meaning of the nuclear revolution: Statecraft and the prospect of Armageddon*. Ithaca, NY: Cornell University Press.

Jo, Dong-Joon, and Erik Gartzke. 2007. Determinants of nuclear weapons proliferation. *Journal of Conflict Resolution* 51 (1): 167.

Johnston, Alastair Iain. 1995. China's new "old thinking": The concept of limited deterrence. *International Security* 20 (3): 5–42.

Jones, Archer, and Andrew J. Keogh. 1985. The Dreadnought revolution: Another look. *Military Affairs* 49 (3): 124–31.

Jones, Clive. 2003. "One size fits all": Israel, intelligence, and the Al-Aqsa Intifada. *Studies in Conflict and Terrorism* 26 (4): 273–88.

Kagan, Frederick, W. 2006. *Finding the target: The transformation of American military policy*. New York: Encounter Book.

Kagan, Frederick, W., and Michael E. O'Hanlon. 2007. Increasing the size and power of the U.S. military. In *Opportunity 08: Independent ideas for our next president*. Washington, DC: Brookings Institution. Available at http://www.brookings.edu/views/papers/ohanlon/20070228kagan.htm.

Kalyvas, Stathis, and Ignacio Sánchez-Cuenca. 2005. Killing without dying: The absence of suicide missions. In Gambetta 2005b, 209–32.

Karau, Mark. 2003. Twisting the dragon's tail: The Zeebrugge and Ostend raids of 1918. *Journal of Military History* 67 (2): 455–81.

Kawakami, Kiyoshi K. 1932. *Japan speaks on the Sino-Japanese crisis*. New York: Macmillan Company.

Kennedy, Paul M. 1976. *The rise and fall of British naval mastery*. New York: Scribner.

Kernan, Alvin B. 2005. *The unknown battle of Midway: The destruction of the American torpedo squadrons*. New Haven, CT: Yale University Press.

Khrushchev, Nikita. 1974. *Khrushchev remembers: The last testament*. Trans. Strobe Talbott. Boston: Little, Brown.

Kier, Elizabeth. 1997. *Imagining war: French and British military doctrine between the wars*. Princeton, NJ: Princeton University Press.

King, Gary, Michael Tomz, and Jason Wittenberg. 2000. Making the most of statistical analyses: Improving interpretation and presentation. *American Journal of Political Science* 44 (2): 347–61.

Kintner, William R., and Harriet Fast Scott. 1968. *The nuclear revolution in Soviet military affairs*. 1st ed. Norman: University of Oklahoma Press.

Kohli, Rajeev, Donald R. Lehmann, and Jae H. Pae. 1999. Extent and impact of incubation time in new product diffusion. *Journal of Product Innovation Management* 16 (2): 134–44.

Kojukharov, Asen N. 1997. In retrospect: The employment of antiship missiles. *Naval War College Review* 50 (4): 118–24.

Kondaki, Christopher D. 2001. Suicide terrorism, an age-old weapon, adds technology. *Defense and Foreign Affairs' Strategic Policy* 29 (September): 8–9.

Kopstein, Jeffrey S., and David A. Reilly. 2000. Geographic diffusion and the transformation of the postcommunist world. *World Politics* 53 (1): 1–37.

Kosiak, Steven M. 2004. Matching resources with requirements: Options for modernizing the US Air Force. Washington, DC: Center for Strategic and Budgetary Assessments.

———. 2006. Analysis of the FY2007 defense budget request. Center for Strategic and Budgetary Assessments. Available at http://www.csbaonline.org/4Publications/Archive/R.20060425.FY07Bud/R.20060425.FY07Bud.pdf.

Kramer, Martin S. 1990. The moral logic of Hizballah. In *Origins of terrorism: Psychologies, ideologies, theologies, states of mind*, ed. Walter Reich, 131–57. Cambridge: Cambridge University Press.

———. 1991. Sacrifice and "self-martyrdom" in Shi'ite Lebanon. *Terrorism and Political Violence* 3 (3): 30–47. Available at http://www.geocities.com/martinkramerorg/Sacrifice.htm.

———. 1993. Hizbullah: The calculus of jihad. In *Fundamentalisms and the state: Remaking polities, economies, and militance*, ed. Martin E. Marty and R. Scott Appleby, 539–55. Chicago: University of Chicago Press.

Krepinevich, Andrew F. 1986. *The army and Vietnam*. Baltimore: Johns Hopkins University Press.

———. 1994. Cavalry to computer: The pattern of military revolutions. *National Interest* 37:30.

———. 2002. *The military-technical revolution: A preliminary assessment* (1992). Washington, DC: Center for Strategic and Budgetary Assessments.

Krueger, Alan B., and Jitka Maleckova. 2003. Education, poverty, and terrorism: Is there a causal connection? *Journal of Economic Perspectives* 17 (4): 119–44.

Kugler, Jacek, and Douglas Lemke, eds. 1996. *Parity and war: Evaluations and extensions of the war ledger*. Ann Arbor: University of Michigan Press.

Kurth, Ronald. 2005. Gorshkov's gambit. *Journal of Strategic Studies* 28 (2): 261–80.

Lake, David A., and Robert Powell, eds. 1999. *Strategic choice and international relations*. Princeton, NJ: Princeton University Press.

Lambert, Nicholas A. 1995a. Admiral Sir John Fisher and the concept of flotilla defence, 1904–1909. *Journal of Military History* 59 (4): 639–60.

———. 1995b. British naval policy, 1913–1914: Financial limitation and strategic revolution. *Journal of Modern History* 67 (3): 595–626.

———. 1999. *Sir John Fisher's naval revolution*. Columbia: University of South Carolina Press.

———. 2001. *The submarine service, 1900–1918*. Publications of the Navy Records Society, vol. 142. Brookfield, VT: Ashgate for the Navy Records Society.

———. 2004. Transformation and technology in the Fisher era: the impact of the communications revolution. *Journal of Strategic Studies* 27 (2): 272–97.

———. 2005. Strategic command and control for maneuver warfare: Creation of the Royal Navy's "war room" system, 1905–1915. *Journal of Military History* 69: 361–410.

Layman, Richard D. 1979. *To ascend from a floating base: Shipboard aeronautics and aviation, 1783–1914*. Rutherford, NJ: Fairleigh Dickinson University Press.

Leeds, Brett Ashley. 2003. Do alliances deter aggression? The influence of military alliances on the initiation of militarized interstate disputes. *American Journal of Political Science* 47 (3): 427–39.

Leeds, Brett Ashely, Andrew G. Long, and Sara McLaughlin Mitchell. 2000. Reevaluating alliance reliability: Specific threats, specific promises. *Journal of Conflict Resolution* 44 (5): 686–99.

Lester, David, Bijou Yang, and Mark Lindsay. 2004. Suicide bombers: Are psychological profiles possible? *Studies in Conflict and Terrorism* 27 (4): 283–95.

Lev, Baruch, and Suresh Radhakrishnan. 2003. The measurement of firm-specific organization capital. *National Bureau of Economic Research Working Paper* W9581 (March). Available at http://papers.ssrn.com/sol3/papers.cfm?abstract_id=389452.

———. 2005. The valuation of organizational capital. In *Measuring capital in the new economy*, ed. Carol Corrado, John Haltiwanger, and Daniel Sichel, 73–110. Chicago: University of Chicago Press.

Levitt, Matthew. 2004. Hamas from cradle to grave. *Middle East Quarterly* 11 (1). Available at http://www.meforum.org/582/hamas-from-cradle-to-grave.

Levy, Jack S. 1981. Alliance formation and war behavior: An analysis of the great powers, 1495–1975. *Journal of Conflict Resolution* 25 (4): 581–613.

Lewis, Jeffrey. 2007. *The minimum means of reprisal: China's search for security in the nuclear age*. Cambridge, MA: MIT Press.

Lieber, Keir A. 2005. *War and the engineers: The primacy of politics over technology*. Ithaca, NY: Cornell University Press.

Lieber, Kier A., and Daryl G. Press. 2006a. The end of MAD? The nuclear dimension of U.S. primacy. *International Security* 30 (4): 7–44.

———. 2006b. The rise of US nuclear primacy. *Foreign Affairs* 85 (2): 42–54.

Lieberman, Marvin B., and David B. Montgomery. 1988. First-mover advantages. *Strategic Management Journal* 9 (1): 41–58.

Luttwak, Edward. 2005. A brief note on "fourth-generation warfare." *Contemporary Security Policy* 26 (2): 227–28.

Mack, Andrew. 1996. Japan and the bomb: A cause for concern? *Asia-Pacific Magazine* (3): 5–9.

Mahajan, Vijay, Eitan Muller, and Frank M. Bass. 1990. New product diffusion models in marketing: A review and directions for research. *Journal of Marketing* 54 (1): 1–26.

Mahnken, Thomas G. 2002. *Uncovering ways of war: US intelligence and foreign military innovation, 1918–1941*. Ithaca, NY: Cornell University Press.

———. 2003. Beyond blitzkrieg: Allied responses to combined-arms armored warfare during World War II. In Goldman and Eliason 2003a, 243–66.

Mandelbaum, Michael. 1981. *The nuclear revolution: International politics before and after Hiroshima*. New York: Cambridge University Press.

———. 1995. Lessons of the next nuclear war. *Foreign Affairs* 74 (2): 22–37.

Marder, Arthur Jacob. 1961. *From the Dreadnought to Scapa Flow: The Royal Navy in the Fisher era, 1904–1919*. Vol. 1. London: Oxford University Press.

————. 1964. *The anatomy of British sea power: A history of British naval policy in the pre-Dreadnought era, 1880–1905.* Hamden, CT: Archon Books.

Marder, Arthur Jacob, Mark Jacobsen, and John Horsfield. 1981. *Old friends, new enemies: The Royal Navy and the Imperial Japanese Navy.* 2 vols. Oxford: Oxford University Press.

Marraro, Howard R. 1942. Unpublished American documents on the naval battle of Lissa (1866). *Journal of Modern History* 14 (3): 342–56.

Marshall, Andrew W. 1995. Revolutions in military affairs: Testimony before the Senate Armed Services Committee, Subcommittee on Acquisition and Technology.

Marshall, Mony G., and Keith Jaggers. 2002. Polity IV project: Political regime characteristics and transitions, 1800–2000. In *University of Maryland's Center for International Development and Conflict Management (CIDCM).* Updated version available at http://www.systemicpeace.org/polity/polity4.htm.

Martin, Thomas G. 1970. A Soviet carrier on the horizon? *Proceedings of the United States Naval Institute* 96:47–51.

Massie, Robert K. 1991. *Dreadnought: Britain, Germany, and the coming of the great war.* New York: Random House.

May, Ernest R. 2000. *Strange victory: Hitler's conquest of France.* New York: Hill and Wang.

Mazarr, Michael J. 2008. The folly of "asymmetric war." *Washington Quarterly* 31 (3): 33–53.

McAllister, Brad. 2004. Al Qaeda and the innovative firm: Demythologizing the network. *Studies in Conflict and Terrorism* 27 (4): 297–319.

McCauley, Clark. 1991. Terrorism, research, and public policy: An overview. In *Terrorism research and public policy*, ed. Clark McCauley. Portland, OR: Frank Cass.

McKitrick, Myra S. 2003. Submarines: Weapons of choice in future warfare. In *Naval strike forum.* Washington, DC: Lexington Institute.

McLean, Iain. 2000. Review article: The divided legacy of Mancur Olson. *British Journal of Political Science* 30 (4): 651–68.

McMaster, H. R. 2008. On war: Lessons to be learned. *Survival* 50 (1): 19–30.

Mearsheimer, John J. 1984. Nuclear weapons and deterrence in Europe. *International Security* 9 (3): 19–46.

————. 1994. The false promise of international institutions. *International Security* 19 (3): 5–49.

Melhorn, Charles M. 1974. *Two-block fox: The rise of the aircraft carrier, 1911–1929.* Annapolis, MD: Naval Institute Press.

Merari, Ariel. 1990. The readiness to kill and die: Suicidal terrorism in the Middle East. In *Origins of terrorism: Psychologies, ideologies, theologies, states of mind*, ed. Walter Reich, 192–210. Cambridge: Cambridge University Press.

Midlarsky, Manus I., Martha Crenshaw, and Fumihiko Yoshida. 1980. Why violence spreads: The contagion of international terrorism. *International Studies Quarterly* 24 (2): 262–98.

Miller, Gene E. 1978. An evaluation of the Soviet Navy. *Proceedings of the Academy of Political Science* 33 (1): 47–56.

Milner, Helen V., and David B. Yoffie. 1989. Between free trade and protectionism: Strategic trade policy and a theory of corporate trade demands. *International Organization* 43 (2): 239–72.

Mintrom, Michael, and Sandra Vergari. 1998. Policy networks and innovation diffusion: The case of state education reforms. *Journal of Politics* 60 (1): 126–48.

Moghadam, Assaf. 2006. Suicide terrorism, occupation, and the globalization of martyrdom: A critique of dying to win. *Studies in Conflict and Terrorism* 29 (8): 707–29.

Moravcsik, Andrew. 1997. Taking preferences seriously: A liberal theory of international politics. *International Organization* 51 (4): 513–53.

Morgan, T. Clifton, and Glenn Palmer. 2003. To protect and to serve: Alliances and foreign policy portfolios. *Journal of Conflict Resolution* 47 (2): 180.

Morison, Samuel Eliot. 1949. *Coral Sea, Midway, and submarine actions, May 1942– August 1942*. Vol. 4, *History of United States naval operations in World War II*. 1st ed. Boston: Little, Brown.

Mueller, John E. 1999a. Dueling counterfactuals. In *Cold war statesmen confront the bomb: Nuclear diplomacy since 1945*, ed. John Lewis Gaddis, Philip H. Gordon, Ernest R. May, and Jonathan Rosenberg, 272–83. London: Oxford University Press.

———. 1999b. The escalating irrelevance of nuclear weapons. In *The absolute weapon revisited: Nuclear arms and the emerging international order*, ed. T. V. Paul, Richard J. Harknett, and James J. Wirtz, 73–98. Ann Arbor: University of Michigan Press.

Mulvenon, James C. 2006. *Chinese responses to U.S. military transformation and implications for the Department of Defense*. Santa Monica, CA: RAND.

Murphy, John F. 2004. The IRA and the FARC in Colombia. *International Journal of Intelligence and CounterIntelligence* 18 (1): 76–88.

Murray, Michelle. 2009. *Identity, insecurity, and great power politics: The tragedy of German naval ambition before the First World War*. Unpublished manuscript, University of Chicago.

Mykytyn, Kathleen, Bijoy Bordoloi, Vicki McKinney, and Kakoli Bandyopadhyay. 2002. The role of software patents in sustaining IT-enabled competitive advantage: A call for research. *Journal of Strategic Information Systems* 11 (1): 59–82.

Nagl, John A. 2005. *Learning to eat soup with a knife: Counterinsurgency lessons from Malaya and Vietnam*. Chicago: University of Chicago Press.

———. 2009. Let's win the wars we're in. *Joint Force Quarterly* 52 (1): 20–26.

Narayan Swamy, M. R. 1994. *Tigers of Lanka, from boys to guerrillas*. New Delhi: Konark Publishers.

National Commission on Terrorist Attacks upon the United States. 2004. *The 9/11 Commission report: Final report of the National Commission on Terrorist Attacks upon the United States*. New York: W. W. Norton.

Natural Resources Defense Council. 2007. *Table of global nuclear weapons stockpiles, 1945–2002*. NRDC Nuclear Program. Available at http://www.nrdc.org/nuclear/nudb/datab19.asp (accessed January 30, 2007).

Naval Aviation News. 1971. Wildcat. *Naval Aviation News*, December. 20–21.

Nimitz, Chester. 1942. Battle of Midway: 4–7 June 1942, online action reports: Commander in chief, Pacific fleet, serial 01849 of 28 June 1942. In *World War II action reports*. College Park, MD: Modern Military Branch, National Archives and Records Administration. Available at http://www.history.navy.mil/docs/wwii/mid1.htm.

Nish, Ian. 1996. Japan and sea power. In *Naval power in the twentieth century*, ed. Nicholas A. M. Rodger, 77–87. London: Macmillan Press Ltd.

Nizamani, Haider K. 2000. *The roots of rhetoric: Politics of nuclear weapons in India and Pakistan.* Westport, CT: Praeger.

Norris, Robert S. 1994. *British, French, and Chinese nuclear forces: Implications for arms control and nonproliferation.* College Park: University of Maryland at College Park.

North, Douglass C. 1981. *Structure and change in economic history.* New York: W. W. Norton.

O'Ballance, Edgar. 1981. *Terror in Ireland: The heritage of hate.* Novato, CA: Presidio Press.

O'Callaghan, Sean. 1998. *The informer.* London: Bantam.

Office of Naval Intelligence. 1947. *The Japanese story of the battle of Midway: A translation.* Washington, DC: U.S. Government Printing Office.

Office of the Chief of Naval Operations. 1947. *U.S. naval aviation in the Pacific.* Washington, DC: Office of the Chief of Naval Operations.

———. 1985. *Understanding Soviet naval developments.* 5th ed. Washington, DC: Office of the Chief of Naval Operations.

Office of the Secretary of Defense. 2005. Unmanned aircraft systems roadmap, 2005–2030. Washington, DC: Department of Defense.

Officer, Lawrence H., and Samuel H. Williamson. 2006. Computing "real value" over time with a conversion between U.K. pounds and U.S. dollars, 1830–2005. Measuring Worth.com. Available at http://www.measuringworth.com/exchange/.

O'Hanlon, Michael E. 2000. *Technological change and the future of warfare.* Washington, DC: Brookings Institution Press.

Olson, Mancur. 1982. *The rise and decline of nations: Economic growth, stagflation, and social rigidities.* New Haven, CT: Yale University Press.

Olukoya, Sam. 2002. Nigeria grapples with e-mail scams. *BBC News*, April 23.

Organski, A.F.K., and Jacek Kugler. 1980. *The war ledger.* Chicago: University of Chicago Press.

O'Rourke, Ronald. 2005. China naval modernization: Implications for US Navy capabilities—background and issues for Congress. Congressional Research Service. November 18.

Pack, S.W.C. 1971. *Sea power in the Mediterranean: A study of the struggle for sea power in the Mediterranean from the seventeenth century to the present day.* London: Barker.

Pape, Robert A. 2003. The strategic logic of suicide terrorism. *American Political Science Review* 97 (3): 343–61.

————. 2005. *Dying to win: The strategic logic of suicide terrorism.* 1st ed. New York: Random House.

Parker, Geoffrey. 1996. *The military revolution: Military innovation and the rise of the west, 1500–1800.* 2nd ed. New York: Cambridge University Press.

Parkes, Oscar. 1970. *British battleships, "Warrior" 1860 to "Vanguard" 1950: A history of design construction and armament.* New and rev. ed. London: Seeley Service.

Parshall, Jonathan B., and Anthony P. Tully. 2005. *Shattered sword: The untold story of the battle of Midway.* 1st ed. Washington, DC: Potomac Books.

Peattie, Mark R. 2001. *Sunburst: The rise of the Japanese naval air power, 1909–1941.* Annapolis, MD: Naval Institute Press.

Pedahzur, Ami. 2005. *Suicide terrorism.* Cambridge, MA: Polity.

Perkovich, George. 1999. *India's nuclear bomb: The impact on global proliferation.* Berkeley: University of California Press.

Perry, John C. 1966. Great Britain and the emergence of Japan as a naval power. *Monumenta Nipponica* 21 (3–4): 305–21.

Pierce, Terry C. 2004. *Warfighting and disruptive technologies: Disguising innovation.* New York: Frank Cass.

Pike, John. 2005a. B-2 production. GlobalSecurity.org. Available at http://www.globalsecurity.org/wmd/systems/b-2-production.htm.

————. 2005b. B-29 Superfortress. GlobalSecurity.org. Available at http://www.globalsecurity.org/wmd/systems/b-29.htm.

————. 2006. R-11 Vikrant. GlobalSecurity.org. Available at http://www.globalsecurity.org/military/world/india/r-vikrant.htm.

Pillsbury, Michael. 2001. China's military strategy toward the U.S.: A view from open sources. Paper commissioned for the U.S.-China Economic and Security Review Commission, Washington, DC.

Pollack, Kenneth M. 2002. *Arabs at war: Military effectiveness, 1948–1991.* Lincoln: University of Nebraska Press.

Polmar, Norman. 1983. *Guide to the Soviet Navy.* 3rd ed. Annapolis, MD: Naval Institute Press.

Posen, Barry R. 1984. *The sources of military doctrine: France, Britain, and Germany between the world wars.* Ithaca, NY: Cornell University Press.

————. 1993. Nationalism, the mass army, and military power. *International Security* 18 (2): 80–124.

Powell, Robert. 1999. *In the shadow of power: States and strategies in international politics.* Princeton, NJ: Princeton University Press.

Prange, Gordon W., Donald M. Goldstein, and Katherine V. Dillon. 1982. *Miracle at Midway.* New York: McGraw-Hill.

Ramasubramanian, R. 2004. Suicide terrorism in Sri Lanka. In *ICPS research papers #5.* New Delhi: Institute of Peace and Conflict Studies. Available at http://www.ipcs.org/IRP05.pdf.

Reiter, Dan, and Allan C. Stam. 2002. *Democracies at war.* Princeton, NJ: Princeton University Press.

Resende-Santos, João. 2007. *Neorealism, states, and the modern mass army*. New York: Cambridge University Press.

Reynolds, Clark G. 1978. *The fast carriers: The forging of an air navy*. Huntington, NY: R. E. Krieger.

Rhodes, Richard. 1995. *Dark sun: The making of the hydrogen bomb*. New York: Simon and Schuster.

Ricolfi, Luca. 2005. Palestinians, 1981–2003. In Gambetta 2005b, 77–130.

Rochlin, Gene I., Todd R. La Porte, and Karlene H. Roberts. 1987. The self-designing high-reliability organization: Aircraft carrier flight operations at sea. *Naval War College Review* 40 (4): 76–90.

Rogers, Everett M. 2003. *Diffusion of innovations*. 5th ed. New York: Free Press.

Rogowski, Ronald. 1983. Structure, growth, and power: Three rationalist accounts. *International Organization* 37 (4): 713–38.

Roland, Alexander. 1997. Technology and war. Paper presented at the Summary Conference on the Study of War, Durham, NC, June.

Ropp, Theodore. 1987. *The development of a modern navy: French naval policy, 1871–1904*. Ed. Stephen S. Roberts. Annapolis, MD: Naval Institute Press.

Rosen, Stephen P. 1991. *Winning the next war: Innovation and the modern military*. Ithaca, NY: Cornell University Press.

———. 2006. After proliferation: What to do if more states go nuclear. *Foreign Affairs* 85 (5): 9–14.

Roskill, Stephen W. 1968. *Naval policy between the wars*. London: Collins.

Ross, Robert S. 2009. China's naval nationalism and the politics of U.S.-China maritime cooperation. *International Security* 34 (2): 46–81.

Sadkovich, James J. 1994. *The Italian Navy in World War II*. Westport, CT: Greenwood Press.

Sagan, Scott D. 1996. Why do states build nuclear weapons? Three models in search of a bomb. *International Security* 21 (3): 54–86.

Sagan, Scott D., and Kenneth N. Waltz. 1995. *The spread of nuclear weapons: A debate*. 1st ed. New York: W. W. Norton.

Sageman, Marc. 2004. *Understanding terror networks*. Philadelphia: University of Pennsylvania Press.

Sankei Shimbun. 2007. Internal government document concludes that it would take Japan over three years to make a prototype model nuclear warhead at cost of 200–300 billion yen. December 25. Translation available at http://www.armscontrolwonk.com/1338/the-sankei-article-on-japanese-nukes.

Sapolsky, Harvey M. 1972. *The Polaris system development: Bureaucratic and programmatic success in government*. Cambridge, MA: Harvard University Press.

———. 1996. The interservice competition solution. *Breakthroughs* 5 (1): 1–4.

Sarkees, Meredith R. 2000. The Correlates of War data on war: An update to 1997. *Conflict Management and Peace Science* 18 (1): 123–44.

Sattinger, Michael. 2004. Capital intensity, neutral technological change, and earnings inequality. *Journal of Income Distribution* 12 (1): 6–20.

Schelling, Thomas C. 1960. *The strategy of conflict.* Cambridge, MA: Harvard University Press.

———. 1966. *Arms and influence.* New Haven, CT: Yale University Press.

Schneider, Barry R. 1994. Nuclear proliferation and counter-proliferation: Policy issues and debates. *Mershon International Studies Review* 38 (2): 209–34.

Schroeder, Paul. 1994. Historical reality vs. neo-realist theory. *International Security* 19 (1): 108–48.

Schultz, Kenneth A. 1999. Do democratic institutions constrain or inform? Contrasting two institutional perspectives on democracy and war. *International Organization* 53 (2): 233–66.

Schumpeter, Joseph A. 1942. *Capitalism, socialism, and democracy.* New York: Harper and Brothers.

Schwartz, Stephen I. 1998. *Atomic audit: The costs and consequences of U.S. nuclear weapons since 1940.* Washington, DC: Brookings Institution Press.

Schweitzer, Yoram. 2002. Suicide terrorism: Development and characteristics. Lecture presented at the International Conference on Countering Suicide Terrorism. Available at http://ftp.beitberl.ac.il/~bbsite/misc/ezer_anglit/klali/01_46.doc.

Selborne Papers. The papers of the 2nd earl of Selborne. Oxford: Bodleian Library.

Siegel, Adam S. 1991. The use of naval forces in the post-war era: U.S. Navy and U.S. Marine Corps crisis response activity, 1946–1990. Alexandria, VA: Center for Naval Analysis. Available at http://www.history.navy.mil/library/online/forces_cold.htm.

Signorino, Curtis S., and Jeffrey M. Ritter. 1999. Tau-B or not Tau-B: Measuring the similarity of foreign policy positions. *International Studies Quarterly* 43 (1): 115–44.

Signorino, Curtis S., and Ahmer Tarar. 2006. A unified theory and test of extended immediate deterrence. *American Journal of Political Science* 50 (3): 586–605.

Simmons, Beth A., and Zachary Elkins. 2004. The globalization of liberalization: Policy diffusion in the international political economy. *American Political Science Review* 98 (1): 171–89.

Simowitz, Roslyn. 1998. Evaluating conflict research on the diffusion of war. *Journal of Peace Research* 35 (2): 211–30.

Singer, J. David. 1987. Reconstructing the correlates of war dataset on material capabilities of states, 1816–1985. *International Interactions* 14:115–32.

Singer, J. David, Stuart Bremer, and John Stuckey. 1972. Capability distribution, uncertainty, and major power war, 1820–1965. In *Peace, war, and numbers,* ed. Bruce Russett, 19–48. Beverly Hills: Sage Publications.

Singer, Peter W. 2009. *Wired for war: The robotics revolution and conflict in the twenty-first century.* New York: Penguin Press.

Singh, Sonali, and Christopher R. Way. 2004. The correlates of nuclear proliferation: A quantitative test. *Journal of Conflict Resolution* 48 (6): 859–85.

Siverson, Randolph M., and Harvey Starr. 1991. *The diffusion of war: A study of opportunity and willingness.* Ann Arbor: University of Michigan Press.

Smith, Alastair, and Allan C. Stam. 2004. Bargaining and the nature of war. *Journal of Conflict Resolution* 48 (6): 783–813.

Snyder, Glenn Herald. 1961. *Deterrence and defense: Toward a theory of national security.* Princeton, NJ: Princeton University Press.

Snyder, Jack L., and Thomas J. Christensen. 1990. Chain gangs and passed bucks: Predicting alliance patterns in multipolarity. *International Organization* 44 (2): 137–68.

Sondhaus, Lawrence. 1992. Strategy, tactics, and the politics of penury: The Austro-Hungarian Navy and the Jeune Ecole. *Journal of Military History* 56 (4): 587–602.

———. 1994. *The naval policy of Austria-Hungary, 1867–1918: Navalism, industrial development, and the politics of dualism.* West Lafayette, IN: Purdue University Press.

———. 2004. *Navies in modern world history: Globalities.* London: Reaktion.

Soule, Sarah A. 1999. The diffusion of an unsuccessful innovation. *Annals of the American Academy of Political and Social Science* 566 (1): 120–31.

Spector, Leonard S. 1992. Repentant nuclear proliferants. *Foreign Policy* (88): 21–37.

Stålenheim, Petter, Elisabeth Sköns, Catalina Perdomo, and Sam Perlo-Freeman, eds. 2006. Military expenditures. In *SIPRI yearbook 2006: Armaments, disarmament, and international security,* 295–324. Oxford: Oxford University Press.

Stam, Allan C. 1996. *Win, lose, or draw: Domestic politics and the crucible of war.* Ann Arbor: University of Michigan Press.

Starr, Harvey, and Benjamin A. Most. 1983. Contagion and border effects on contemporary African conflict. *Comparative Political Studies* 16 (1): 92.

Starr, Harvey, and Randolph M. Siverson. 1998. Cumulation, evaluation, and the research process: Investigating the diffusion of conflict. *Journal of Peace Research* 35 (2): 231–37.

Steele, Dennis. 2006. Growing pains: Getting army UAV aviation off the ground. *Army* (January): 20–26.

Strang, David. 1991. Adding social structure to diffusion models: An event history framework. *Sociological Methods and Research* 19 (3): 324.

Strang, David, and John W. Meyer. 1993. Institutional conditions for diffusion. *Theory and Society* 22 (4): 487–511.

Strang, David, and Sarah A. Soule. 1998. Diffusion in organizations and social movements: From hybrid corn to poison pills. *Annual Review of Sociology* 24 (1): 265–90.

Stulberg, Adam. 2005. Managing military transformations: Agency, culture, and the U.S. carrier revolution. *Security Studies* 14 (3): 489–528.

Suchman, Mark C., and Dana P. Eyre. 1992. Military procurement as rational myth: Notes on the social construction of weapons proliferation. *Sociological Forum* 7 (1): 137–61.

Sumida, Jon T. 1979. British capital ship design and fire control in the Dreadnought era: Sir John Fisher, Arthur Hungerford Pollen, and the battle cruiser. *Journal of Modern History* 51 (2): 205–30.

———. 1989. *In defence of naval supremacy: Finance, technology, and British naval policy, 1889–1914.* Boston: Unwin Hyman.

———. 1993. British naval operational logistics, 1914–1918. *Journal of Military History* 57 (3): 447–80.

———. 1995. Sir John Fisher and the Dreadnought: The sources of naval mythology. *Journal of Military History* 59 (4): 619–37.

———. 1996. The quest for reach: The development of long-range gunnery in the Royal Navy, 1901–1912. In *Tooling for war: Military transformation in the industrial age: Proceedings of the sixteenth Military History Symposium of the United States Air Force Academy*, ed. Stephen D. Chiabotti. Chicago: Imprint Publications.

———. 2006. Geography, technology, and British Naval strategy in the Dreadnought era. *Naval War College Review* 59 (3): 89–102.

Tannenwald, Nina, and Richard Price. 1996. Norms and deterrence: The nuclear and chemical weapons taboos. In *The culture of national security: Norms and identity in world politics*, ed. Peter J. Katzenstein, 114–52. New York: Columbia University Press.

Tekwani, Shyam. 2006. The LTTE's online network and its implications for regional security. RSIS Working Papers, no. 104. January. Available at http://www.rsis.edu.sg/publications/WorkingPapers/WP104.pdf.

Tellis, Ashley J. 1985. The naval balance in the Indian subcontinent: Demanding missions for the Indian Navy. *Asian Survey* 25 (12): 1186–213.

Tellis, Gerard J., and Peter N. Golder. 2002. *Will and vision: How latecomers grow to dominate markets*. New York: McGraw-Hill.

Terrorism Knowledge Base. 2006. *Memorial Institute for the Prevention of Terrorism*. Available at http://www.start.umd.edu/start/data/tops/ (accessed July 2006).

Thomas, Timothy L. 2000. Kosovo and the current myth of information superiority. *Parameters* 30 (1): 13–29.

Thompson, Loren. 2001. Aircraft carrier (in)vulnerability: What it takes to successfully attack an American aircraft carrier. In *Naval strike forum*. Washington, DC: Lexington Institute.

Till, Geoffrey. 1979. *Air power and the Royal Navy, 1914–1945: A historical survey*. London: Macdonald and Jane's.

———. 1996. Adopting the aircraft carrier: The British, American, and Japanese case studies. In *Military innovation in the interwar period*, ed. Williamson Murray and Allan R. Millett, 191–226. Cambridge: Cambridge University Press.

———. 2005. Holding the bridge in troubled times: The cold war and the navies of Europe. *Journal of Strategic Studies* 28 (2): 309–37.

Toft, Monica D. 2003. *The geography of ethnic violence: Identity, interests, and the indivisibility of territory*. Princeton, NJ: Princeton University Press.

Trachtenberg, Marc. 1988. A "wasting asset": American strategy and the shifting nuclear balance, 1949–1954. *International Security* 13 (3): 5–49.

Tracy, Nicholas, ed. 1997. *The collective naval defence of the empire, 1900–1940*. Vol. 136, Publications of the Navy Records Society. Brookfield, VT: Ashgate for the Navy Records Society.

Tritten, James J., and Luigi Donolo. 1995. *A doctrine reader: The Navies of United States, Great Britain, France, Italy, and Spain*. Newport, RI: Center for Naval Warfare Studies, Naval War College.

True, Jacqui, and Michael Mintrom. 2001. Transnational networks and policy diffusion: The case of gender mainstreaming. *International Studies Quarterly* 45 (1): 27–57.

Tushman, Michael L., and Philip Anderson. 1986. Technological discontinuities and organizational environments. *Administrative Science Quarterly* 31 (3): 439–65.

Tweedmouth Papers. The papers of the 2nd Baron Tweedmouth. Portsmouth, UK: Royal Naval Museum Library/Admiralty Library.

Unger, Brigitte, and Frans Van Waarden. 1999. Interest associations and economic growth: A critique of Mancur Olson's Rise and Decline of Nations. *Review of International Political Economy* 6 (4): 425–67.

United Kingdom and France. 1904. Entente Cordiale: Declaration between the United Kingdom and France respecting Egypt and Morocco, together with the secret articles signed at the same time. Available at http://net.lib.byu.edu/~rdh7/wwi/1914m/entecord.html.

U.S. Air Force. 2006a. B-2 Spirit. U.S. Air Force fact sheet. Available at http://www.af.mil/factsheets/factsheet_print.asp?fsID=82&page=1.

———. 2006b. GBU-39B, Small diameter bomb weapon system. U.S. Air Force fact sheet. Available at http://www.af.mil/factsheets/factsheet_print.asp?fsID=4500&page=1.

———. 2006c. Joint direct attack munitions GBU 31/32/38 U.S. Air Force fact sheet. Available at http://www.af.mil/factsheets/factsheet_print.asp?fsID=108&page=1.

———. 2008. MQ-1 Predator unmanned aircraft system. U.S. Air Force fact sheet. Available at http://www.af.mil/information/factsheets/factsheet.asp?id=122.

U.S. Department of Defense. 2007. *Military power of the People's Republic of China: A report to Congress pursuant to the National Defense Authorization Act fiscal year 2000.* Washington, DC: Office of the Secretary of Defense.

U.S. Department of State. 1973. *Germany and Austria.* Vol. 2, *Foreign relations of the United States of America.* Washington, DC: U.S. Government Printing Office.

———. 1974. *Eastern Europe; the Soviet Union.* Vol. 4, *Foreign relations of the United States of America.* Washington, DC: U.S. Government Printing Office.

van Blyenbrugh, Peter. 2008. Unmanned aircraft systems: The current situation. Brussels: UAS ATM Integration Workshop, EUROCONTROL.

Van Creveld, Martin 2005. It will continue to conquer and spread. *Contemporary Security Policy* 26 (2): 229–32.

Vasquez, John A. 1991. The deterrence myth: Nuclear weapons and the prevention of nuclear war. In *The long postwar peace: Contending explanations and projections,* ed. Charles W. Kegley Jr., 205–23. New York: HarperCollins Publishers.

———. 1995. Why do neighbors fight? Proximity, interaction, or territoriality. *Journal of Peace Research* 32 (3): 277–93.

Vego, Milan N. 1996. *Austro-Hungarian naval policy, 1904–14.* Portland, OR: Frank Cass.

Vernoski, Kenneth. 1986. Soviet naval aircraft. In *The Soviet Navy: Strengths and liabilities,* ed. Bruce W. Watson and Susan M. Watson, 114–24. Boulder, CO: Westview Press.

Vickers, Michael G., and Robert C. Martinage. 2004. *The revolution in war*. Washington, DC: Center for Strategic and Budgetary Assessments.

Walker, Jack L. 1969. The diffusion of innovations among the American states. *American Political Science Review* 63 (3): 880–99.

Walsh, James J. 1997. Surprise down under: The secret history of Australia's nuclear ambitions. *Nonproliferation Review* 5 (1): 1–20.

Walt, Stephen M. 1987. *The origins of alliances*. Ithaca, NY: Cornell University Press.

Waltz, Kenneth N. 1979. *Theory of international politics*. New York: McGraw-Hill.

———. 1995. More may be better. In *The spread of nuclear weapons: A debate*, ed. Scott D. Sagan and Kenneth N. Waltz, 1–46. New York: W. W. Norton.

Ward, Michael D., and Kristian Skrede Gleditsch. 1998. Democratizing for peace. *American Political Science Review* 92 (1): 51–61.

Watson, Bruce W., and Susan M. Watson. 1986. Looking toward the future. In *The Soviet Navy: Strengths and liabilities*, ed. Bruce W. Watson and Susan M. Watson, 289–99. Boulder, CO: Westview Press.

Weber, Max. 2002. *The Protestant ethic and the spirit of capitalism*. 3rd ed. Los Angeles: Roxbury Publishing Company.

Weick, Karl E., and Karlene H. Roberts. 1993. Collective mind in organizations: Heedful interrelating on flight decks. *Administrative Science Quarterly* 38 (3): 357–81.

Weinberg, Leonard, and Ami Pedahzur. 2004. *Religious fundamentalism and political extremism*. Portland, OR: Frank Cass.

Welch, Thomas J. 1999. Revolution in military affairs: One perspective. In *Strength through cooperation: Military forces in the Asian-Pacific Region*, ed. Frances Omori and Mary A. Sommerville. Washington, DC: National Defense University Press.

Wendt, Alexander, and Michael Barnett. 1993. Dependent state formation and third world militarization. *Review of International Studies* 19 (4): 321–47.

Wettern, Desmond. 1982. *The decline of British seapower*. London: Jane's Publishing Company, Ltd.

Weyland, Kurt G. 2004. Neoliberalism and democracy in Latin America: A mixed record. *Latin American Politics and Society* 46 (1): 135–57.

Wildenberg, Thomas. 2005. Sheer luck or better doctrine? *Naval War College Review* 58 (1): 121–35.

Wilkinson, Paul. 1986. *Terrorism and the liberal state*. 2nd ed. New York: New York University Press.

Williamson, Samuel H. 2006. Five ways to compute the relative value of a U.S. dollar amount, 1790–2005. MeasuringWorth.com. Available at http://www.measuring worth.com/uscompare/.

Williamson, Samuel R., and Steven L. Reardon. 1993. *The origins of U.S. nuclear strategy, 1945–1953*. New York: St. Martin's Press.

Wilson, James Q. 1989. *Bureaucracy: What government agencies do and why they do it*. New York: Basic Books.

Wirtz, James. 2005. Politics with guns: A response to T.X. Hammes. *Contemporary Security Policy* 26 (2): 222–26.

World Islamic Front. 1998. Jihad against Jews and crusaders. Available at http://www
.fas.org/irp/world/para/docs/980223-fatwa.htm.

Yegorova, Natalia. 2005. Stalin's conception of maritime power: Revelations from the
Russian archives. *Journal of Strategic Studies* 28 (2): 157–86.

Zarzecki, Thomas W. 2002. *Arms diffusion: The spread of military innovations in the international system.* New York: Routledge.

Zisk, Kimberly Marten. 1993. *Engaging the enemy: Organization theory and Soviet military innovation, 1955–1991.* Princeton, NJ: Princeton University Press.

Zucker, Lynne G., and Michael R. Darby. 1996a. Costly information: Firm transformation, exit, or persistent failure. *American Behavioral Scientist* 39 (8): 959–74.

———. 1996b. Star scientists and institutional transformation: Patterns of invention
and innovation in the formation of the biotechnology industry. *Proceedings of the
National Academy of Sciences of the United States of America* 93:12709–16.

INDEX

Abrahms, Max, 173n8

adoption-capacity theory, ix, 3, 8–12, 30, 53, 225; and capacity vs. interests, 21, 21n8, 25–26, 53, 211; and choices for major powers, 44; and compared to prior research, 21; and comparison with cultural approach, 58–60; and comparison with neorealism, 55–57; and comparison with norms-based approach, 57–58; and first-mover advantages, 49–50; and interest vs. capacity, 53; limitations of, 12, 39, 209, 225; and nonstate actors, 12, 62–63 (*see also* suicide terrorism); potential critiques of, 53–55; and potential utility in information age, 215–16; and power transitions/balance of power, 43–45, 47–48, 210–11, 212, 225; and strategic choices by states, 40–41, 211; and success or failure, 41–42; and system-level diffusion predictions, 39–40; and system-level impact predictions, 45–46; theoretical contribution of, 209–11, 213–14. *See also* battlefleet warfare; carrier warfare; nuclear weapons

Admiral Gorshkov (Soviet ship), 77, 90

Admiral Kuznetsov (Soviet ship), 77, 85

aircraft carrier: defenses of, 83, 83n31; definition of, 68n; diffusion of, 79–80. *See also* carrier warfare

alliances: and carrier case, 87; COW data on, 114; and innovation diffusion, 27, 176n14; and nuclear case, 99, 121–22, 133, 189, 210; as a response to innovations, 26–27, 29; and suicide terrorism case, 187, 188–89

Al Aqsa Intifada, 167, 173

Al Aqsa Martyrs Brigade, 189

Al Qaeda, 10, 166–67, 173, 179, 183–84, 195, 197–201, 206, 211, 223, 224–25; creation of, 198; and critical task focus, 10; and diffusion from Hezbollah, 166–67, 177, 194, 200; and diffusion from the LTTE, 200; organizational capital of, 166, 198–200; and pre-9/11 attacks, 198; and role in diffusion of suicide terrorism, 167, 187, 194; and September 11, 2001 attacks, 198, 201;

and USS *Cole* bombing, 198, 200; and World Trade Center (1993) attack, 201

Al-Zawahiri, Ayman, 199–200

Amal, 175, 191

American Memorial Institute for the Prevention of Terrorism, 181–82, 182n28, 201; coding scheme of, 183

Andrea Doria (Italian ship), 93

Andres, Richard B., 43n52

Anglo-German arms race, 136, 152

antiship missiles, 66, 82–83; as counter to carriers, 83

Apple, 6

arms races, 5n, 136, 152

Arromanches (French ship), 77

Arab-Israeli wars, 7, 18

Arreguin-Toft, Ivan, 207

Asal, Victor, 187

assassins, 172

Atran, Scott, 173

Austro-Hungarian Empire: and Napoleonic warfare, 28, 41; and World War I, 11–12, 213

Austro-Hungarian Navy, 159–61; experimentation by, 160; interests of, 159; organizational age of, 160; and response to *Dreadnought*, 149; spending of, 160

Axelrod, Robert, 8n9

B-2 bomber, 31, 102, 217, 218

B-29 Superfortress, 102–3, 129n39

balance of power: impact of diffusion on, ix, 4–5, 11, 23, 47–48. *See also* adoption capacity theory; major military innovations; power transitions

Basque Fatherland and Freedom Group (ETA), ix, 12, 167, 170, 192, 203, 206–7, 211

battle cruiser, 139; and importance for Fisher, 193; speed of, 139

battlefleet warfare, 63–64, 134–65; and Anglo-German arms race, 136; debut of, 138; diffusion of, 147–48; financial intensity of,